H. V. Shurmer
Microwave Semiconductor Devices

H. V. Shurmer, Ph.D., C.Eng., F.I.E.E. is a Senior Research Fellow at Warwick University, England

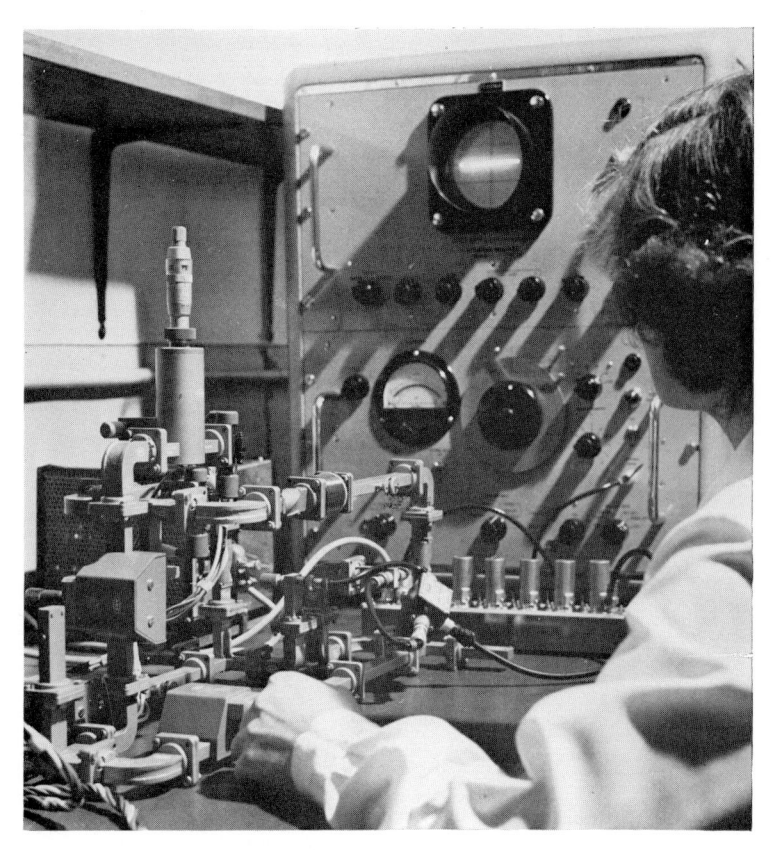

Flicker-noise measuring equipment
(*GEC/AEI*)

Microwave Semiconductor Devices

H. V. SHURMER
PhD, CEng, FIEE

Senior Research Fellow
Warwick University

WILEY-INTERSCIENCE
A Division of John Wiley & Sons, Inc., New York

First published 1971

Published in the U.S.A. by Wiley-Interscience
A Division of John Wiley & Sons, Inc., New York

© H V Shurmer 1971

Library of Congress Catalog Card No: 70-165546

PITMAN ISBN: 0 273 41874 2
WILEY ISBN: 0 471 78990 9

Made in Great Britain at the Pitman Press, Bath

TK 7871.85
S53
1971b

SMU LIBRARY

Preface

The field of microwaves nowadays embraces innumerable types of military and civil radars, communication systems, telemetry, instrumentation, industrial heating, etc., and numerous semiconductor devices have been evolved or are being developed for microwave applications. The level at which these are discussed in the present book is appropriate for students on an engineering degree course or its equivalent. It is hoped, however, that the breadth of the subject-matter may give it an extended usefulness to established engineers in the design and manufacture of semiconductor devices as well as to others with an interest in electronics who are seeking an outline knowledge of this somewhat specialized subject.

The entire microwave area, which now represents a significant part of the electronics industry, is of relatively recent origin, and the influence of semiconductors on its growth has been so important that it is felt worth while in this preface to include some historical notes which will also serve as a background for the variety of devices which are discussed.

Although some experimental work was being conducted in the early 'thirties, microwaves did not assume practical importance until the Second World War, and the first microwave semiconductor device was of course the crystal valve. The impetus for the development of microwaves came from a need to improve upon the accuracy of 1·5 and 10 metre radars, chains of which played so vital a part in the Battle of Britain in 1940. Centimetric radar had, however, to await J. T. Randall's invention of the novel form of magnetron which made available for the first time peak powers measured in hundreds of kilowatts. For detection, attention was immediately turned to the crystal valve, which, by virtue of its low capacitance and short transit time, proved a better rectifier of centimetric radiation than the thermionic valve.

The first crystals for 10 cm radar were being made industrially by a team under T. H. Kinman at the British Thomson-Houston Co. (now incorporated in G.E.C./A.E.I.) towards the end of 1940, and it was the 10 cm radars as blind-bombing and navigational aids which, in 1943, made possible an Allied victory in the Battle of the Atlantic against German U-boats. However, the history of the crystal valve goes back almost to the turn of the century. The first patent on a crystal detector was filed in America in March, 1906, by Dunwoody, a retired Brigadier General of the United States Army, but the roots of the subject were formed earlier. In 1894 Sir Oliver Lodge gave the name "coherer" to a detecting device reported four years before by Professor Branley of Paris, who first observed that an electric spark had the power of changing the conductivity of loose masses of powdered conductors. Knowledge pertaining to devices for detecting electric waves has been traced as far back as 1835, to Munk of Rosenschoeld, who clearly described the permanent increase in electrical conductivity of a mixture of tin filings, carbon and other conductors resulting from the passage through it of the discharge of a Leyden jar.

Crystal valves have undergone fairly continuous development since 1940, silicon and germanium types being still widely used as detectors and mixers. Work on semiconductor materials was given impetus by the advent of point-contact crystals, and the high-purity germanium which resulted was an essential ingredient in the invention of the transistor by Bardeen, Brattain and Shockley in 1948 at the Bell Telephone Laboratories. Ten years later transistor technology was beginning to repay its debt to the microwave art, and a classic paper by Uhlir in 1958 described the use of graded p–n junctions to make a new family of components having microwave potential, including low-noise amplifiers, frequency converters, harmonic generators, limiters and switches.

During the present decade all the components which Uhlir outlined (employing varactors and p–i–n diodes) have indeed become important in microwave engineering. To the original devices, however, must now be added several others. First there is the tunnel diode, utilizing a negative-resistance effect and invented by Esaki in 1957, whose application to microwaves was first described by Sommers in 1959. A device dependent on incipient tunnelling, the backward diode, resulted from the tunnel diode, and the application of this to microwave mixing and detection was reported by Eng in 1961.

Another proposal for a negative-resistance diode with microwave potential was made by W. T. Read in 1958, but it was not until 1965 that a modified form of this device was reported by Johnston, De Loach and Cohen to have given oscillations at microwave frequencies. By the end of 1966 Misawa had obtained more than a watt of continuous power by this means. A prediction of yet another negative-resistance phenomenon was made by Hilsum in 1962, i.e. transferred electron oscillation, which was to

Preface

explain the observation by Gunn in 1963 of microwave oscillations in homogeneous specimens of gallium arsenide at a critical electric field. In 1965 a team led by Hilsum obtained c.w. microwave radiation from epitaxially grown films of gallium arsenide, and in the following year Petzinger, Hahn and Matzell announced a three-terminal device based on the Gunn effect.

In 1962 the charge-storage diode was first described by Moll, Krakauer and Shen, and a new name "step recovery diode" (later "snap-off diode") was subsequently coined by Krakauer for a sub-group of these diodes with potential as high-order harmonic generators of microwaves. By 1965 these diodes were reported by R. D. Hall to be giving about 20 mW of c.w. power in the 3 cm band. By 1967, C. B. Swann had obtained nearly 5 W at a wavelength close to 2 cm.

Metal-semiconductor diodes of the Schottky-barrier type were described by Kahng in 1963, and by 1965 these had also become important in the microwave field, when H. F. Cooke reported that as mixers they were giving better noise figures than any previous point-contact diodes. In the same paper he reported also that transistor oscillators had been increased in frequency to give fundamental power output at 5 GHz, and by the following year harmonic power from transistors was available in the 3 cm band. Yet another device arrived in 1965—the Mizushima diode—resulting from a fast switching effect observed in thin films of gallium arsenide, suggesting potential applications with strip transmission lines.

At the present time work on components and sub-systems employing integrated microwave devices in strip line is being actively pursued in many laboratories, and developments in this direction are having a profound effect on devices applications. Nevertheless, the basic principles of the devices herein described will undoubtedly survive such a trend, even though their embodiments may become radically transformed.

H.V.S.

Acknowledgements

The author is indebted to innumerable colleagues and friends for their assistance and criticism in the writing of this book. The following in particular have helped on technical aspects: T. H. B. Baker, J. C. Bass, I. Bott, P. N. Butcher, J. E. Curran, R. Giblin, J. G. Gissing, R. G. Hibberd, N. R. Howard, T. M. Hyltin, J. S. Lamming, T. E. Oxley and J. R. G. Twissleton.

Thanks are due to Dr J. E. Stanworth, Director of G.E.C./A.E.I. Central Research and to Dr J. Shields, Managing Director of G.E.C. Semiconductors, without whose encouragement the book would not have been written.

Acknowledgement is expressed to the bodies listed below for permission to reproduce copyright material:

H.M. Stationery Office
American Institute of Physics
Institute of Electrical and Electronic Engineers
Institution of Electrical Engineers
Institution of Electronic and Radio Engineers
University of Manchester Institute of Science and Technology

A.E.I. Semiconductors
General Electric Co. (England)
Kemptron International
Sylvania Electric Products
Texas Instruments

G. Bell & Sons
Litton International Publishing Co.
Oxford University Press

Bell System Technical Journal
Electronic Equipment News
Electronic Engineering
Electronic Design
Electronic Technology
Electro-technology
Industrial Electronics
Journal of Electronics and Control
Microwave Journal
Microwaves
New Scientist
Physics Letters

Contents

	Preface	vii
	Acknowledgements	x
1	Point-contact Crystal Diodes	1
2	Varactor Diodes	44
3	Schottky-barrier Diodes	93
4	Tunnel Diodes	109
5	Backward Diodes	124
6	p–i–n Diodes	134
7	Transistors	150
8	Gunn-effect Devices	164
9	Avalanche Diode Oscillators	184
10	Integrated Circuits	203
	Index	219

1

Point-contact Crystal Diodes

1.1 INTRODUCTION

The device most extensively used in the reception of microwave signals is still the point-contact crystal diode, operated either as a mixer in a superheterodyne receiver or as a low-level detector in a video receiver. The basic difference in operation between superheterodyne and video systems is illustrated in Fig. 1.1. For a mixer, the limiting sensitivity per

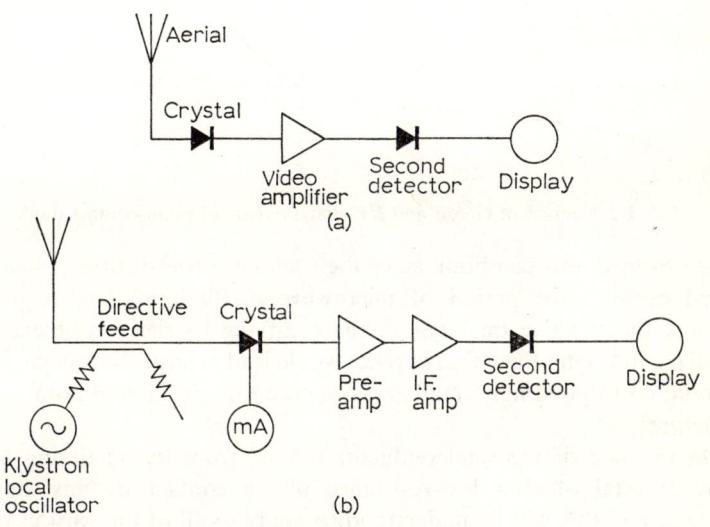

FIG 1.1 Essential elements of microwave receivers
(a) Video receiver   (b) Superheterodyne receiver

unit bandwidth in the presence of noise is inherently greater by several orders of magnitude, a result which is suggested by considering the I/V characteristic illustrated in Fig. 1.2(b).

1

With a detector, the rectified signal depends on the non-linearity of the characteristic over a very small region close to a fixed operating point. In a mixer, however, the operating point is no longer fixed but is driven by the local oscillator excitation between regions of very low and very high differential resistance, and it is this utilization of the full characteristic which greatly enhances the sensitivity.

A point-contact crystal diode thus operates essentially as a resistance varied at a microwave rate. Metal-semiconductor structures are well

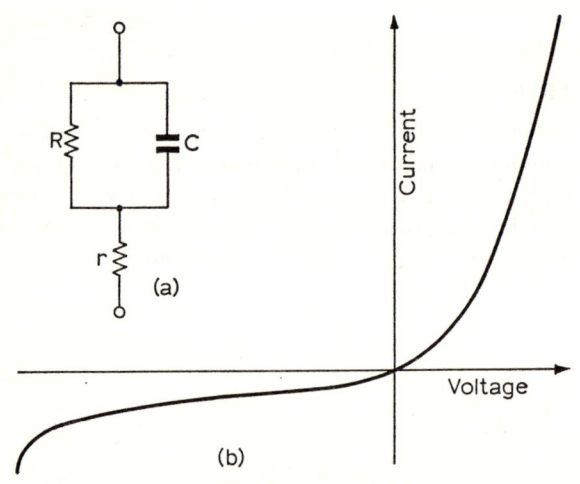

Fig 1.2 Equivalent circuit and I/V characteristic of point-contact diode

suited to meet this condition since they can have transit times small compared even to the period of microwave oscillations. This is possible because the effective thickness of the rectifying barrier can be extremely small ($c.$ 10^{-5} cm). Such thin barriers would lead to large shunt capacitances without correspondingly small areas—hence the adoption of point-contact structures.

On the side of the semiconductor remote from the rectifying barrier there is established a low-resistance ohmic contact by any standard method, and this will be understood to apply to all of the various diodes discussed in this book.

Very small contact areas to the rectifying junction unfortunately lead to constriction of the current flow, creating a spreading resistance r which is effectively in series with the resistance R and capacitance C of the barrier. A simple equivalent circuit for the basic contact element is therefore as shown in Fig. 1.2(a), although it must be noted that both R and C are non-linear functions of voltage.

Silicon has been the semiconductor material principally used in all crystal diodes and exclusively for detectors [1], although in recent years germanium has also been employed with considerable success for mixers [2, 3, 4]. In addition, gallium arsenide has proved particularly interesting at the highest frequencies, but because of its more difficult technology, it has not so far been employed on a commercial scale [5, 6]. Silicon is invariably doped p-type, whilst germanium and gallium arsenide are doped with n-type impurities. The metal whiskers used with silicon are usually of tungsten or a molybdenum-tungsten alloy; titanium is used with germanium; phosphor-bronze is a preferred metal with gallium arsenide.

Surface preparation of the semiconductor, involving heat treatments, mechanical polishing and chemical etching, is always important. The general aims with these processes are to control surface resistivity and surface states, a secondary purpose being to provide a non-skid surface for the metal whisker. It is also customary to "form" the contacts by mechanical shocks (tapping) with silicon diodes, but electrical pulsing techniques are employed with germanium and gallium arsenide. The judicious use of such forming processes improves the I/V characteristic, leading to enhanced conversion efficiency and reduced noise as well as tending to increase mechanical stability.

One of the first applications of the technique of growing thin layers of semiconductors epitaxially on semiconductor substrate materials was to silicon crystals, the object being to improve the relatively high spreading resistance associated with silicon [7]. The conventional form of point contact, assumed to be circular, is shown in Fig. 1.3(a), and the corresponding epitaxial form in Fig. 1.3(b). If the film in (b) has the same resistivity as the thick layer in (a), the spreading resistance is improved in the ratio

$$\frac{r'}{r} = \frac{4}{\pi}\frac{w}{a} \quad \text{if } w \gg a \tag{1.1}$$

Fig. 1.3(c) illustrates the epitaxial case with w/a ratios up to 3·0.

There are practical considerations which limit the improvement which can be achieved, principally transfer of doping material from the substrate during the growth of the thin film, and certainly not all present silicon crystal diodes employ epitaxial material. The technique, moreover, has not been applied with success to germanium microwave diodes.

Microwave Semiconductor Devices

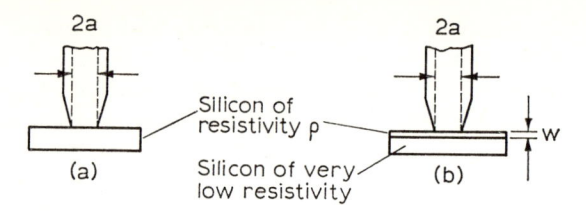

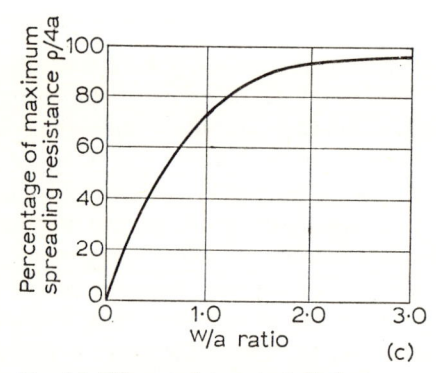

FIG 1.3 Silicon point-contact diodes
(a) Conventional: $r = \rho/4a$
(b) Epitaxial: $r' = w\rho/\pi a^2$
(c) Relation between maximum spreading resistance and w/a ratio
(From Ref. 7)

1.2 SEMICONDUCTOR THEORY

Simple models for the conditions prevailing at metal/semiconductor barriers are illustrated in Fig. 1.4, where

D = Barrier thickness with bias applied
D_0 = Barrier thickness with zero bias
eV_D = Barrier height seen from semiconductor
V_B = Bias voltage
V_D = Diffusion potential
W_C = Bottom of conduction band
W_F = Fermi level
W_V = Top of valence band
ϕ_m = Thermionic work function of metal
ϕ_n = Height of conduction band above Fermi level
ϕ_{ns} = Barrier height seen from metal, n-type semiconductor
ϕ_{ps} = Barrier height seen from metal, p-type semiconductor
χ = Electron affinity (energy required to transfer an electron from bottom of conductor band to free space)

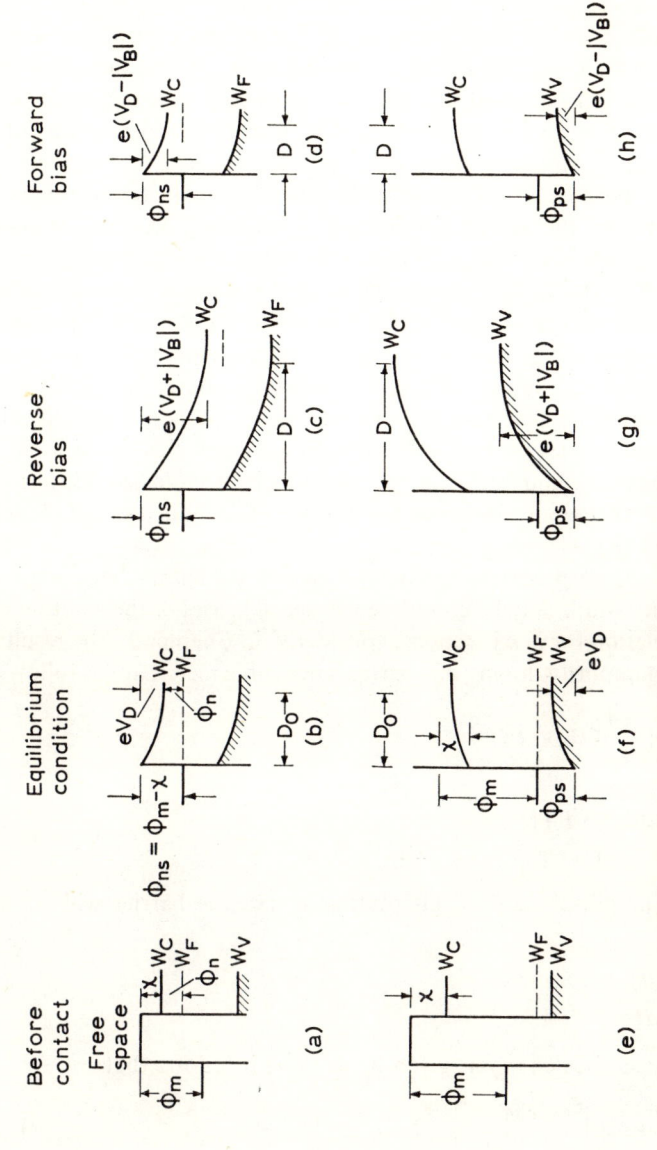

Fig 1.4 Conditions at a simple metal/semiconductor barrier

(a)–(d) n-type semiconductor
(e)–(h) p-type semiconductor

(From Ref. 8)

The energy profile for a metal and separated semiconductor (*n*-type) is shown in Fig. 1.4(*a*). On establishing electronic equilibrium the Fermi levels must become coincident on account of thermodynamic requirements and the profile becomes that shown at (*b*), the Fermi level in the metal having risen by an amount equal to the difference in thermionic work functions. It should be added that for such a model to be applicable it is necessary that $\phi_n < (\phi_m - \chi) \gg kT$, and it is further assumed that the charge due to minority carriers may be neglected [8]. The mobile carriers are removed from the barrier region within the semiconductor by the field which results from the potential difference at the metal interface, leaving a space-charge region of width D_0 which is depleted of carriers.

In Figs. 1.4(*c*) and (*d*) are shown the energy conditions appropriate to forward and reverse bias, respectively, forward current flow corresponding to negative values of V_B, and vice versa. It is here assumed that the space charge within the barrier region is not significantly altered by the current flowing, and that the electronic equilibrium at $x \geqslant D$ is not appreciably disturbed. This implies that outside the barrier region the Fermi level is uniquely defined, although this is not true within it under conditions of current flow. Analogous situations for *p*-type semiconductors are shown in Figs. 1.4(*e*)–(*h*).

The shape of the potential distribution within the space-charge region and the barrier width may be considered as special cases of the solution of Poisson's equation for *p–n* junctions, considered in Chapter 2. The results, for a uniform impurity doping of *n* carriers per cubic metre, are as follows:

$$\phi = V + \frac{en}{2\epsilon}(D - x)^2$$

with

$$D = \left[\frac{2\epsilon(\phi_0 - V)}{en}\right]^{1/2} \tag{1.2}$$

For small-amplitude applied alternating signals the barrier will act as a capacitor:

$$C = A\frac{dq}{dV}$$

and since the surface charge is given by $neD(V)$ it follows that

$$C = \frac{A\epsilon}{D} = A\left[\frac{\epsilon en}{2(\phi_0 - V)}\right]^{1/2} \tag{1.3}$$

It will be noted that the capacitance is the same as that of a parallel-plate capacitor of area A and thickness D.

The I/V relationship may be evaluated by assuming that standard diode theory is applicable. This would appear to require that the barrier is thin compared with the carrier mean free path. This cannot be assumed to hold for silicon diodes, in which the mean free path ranges from 1×10^{-6} to 3×10^{-6} cm. However, more thorough reasoning indicates that the thickness of importance is not that of the total barrier but rather the distance over which the potential energy changes by kT.

In applying the diode theory it is assumed that the electron distribution is given by the Maxwell–Boltzmann law relating to the bulk properties of the semiconductor [9]. It is found that dn, the number of carriers per unit volume crossing unit area per second in the $-x$ direction with velocity between u and $u + du$ is given by

$$dn = n \left(\frac{m}{2\pi kT}\right)^{1/2} e^{-mu^2/2kT} \, u \, du$$

The number of carriers with sufficient kinetic energy W to surmount the barrier is

$$\int_{e(\phi_0 - V)}^{\infty} \left(\frac{dn}{dW}\right) dW = \tfrac{1}{2} n \bar{u} \, e^{-e(\phi_0 - V)/kT}$$

where $\bar{u} = (2kT/\pi m)^{1/2}$ is the average velocity, m being the effective mass of a carrier.

The forward current density, from semiconductor to metal, is then

$$J_1 = \tfrac{1}{2} ne\bar{u} \, e^{-e(\phi_0 - V)/kT}$$

The reverse current density, from metal to semiconductor, is constant in this approximation, and since it must equal J_1 when $V = 0$,

$$J_2 = \tfrac{1}{2} ne\bar{u} \, e^{-e\phi_0/kT}$$

The net current density from semicoductor to metal is then $J = J_1 - J_2$, or

$$J = \tfrac{1}{2} ne\bar{u} \, e^{-e\phi_0/kT} (e^{eV/kT} - 1) \tag{1.4}$$

The current/voltage characteristic is thus predicted to be of the form

$$I = I_0(e^{\alpha V} - 1) \tag{1.5}$$

where V is the voltage across the barrier and

$$\alpha = \frac{e}{kT} \approx 40/\text{V}$$

In practice, point-contact diodes have characteristics which depart appreciably from the ideal form. In particular, the value of α is only about half that suggested above, and the reverse current, far from being constant with voltage, is usually of substantially exponential form.

The factors leading to non-ideal performance are closely allied to the semiconductor surface conditions. In addition to the energy levels appropriate to the bulk semiconductor there are at the surface additional states which can trap charges and alter the barrier height [10]. In practice, these surface states exert a predominating influence on the barrier rather than the difference in work functions between metal and semiconductor. They are influenced by the prior treatment of the semiconductor surface and by the forming conditions.

Since it has been found to reduce reverse current, part of the semiconductor processing is aimed at providing an artificial depletion layer close to the surface, this being achieved through the out-diffusion of impurity atoms during heat treatment. Any such non-uniformity in the impurity distribution will clearly cause the characteristics to depart from those of the model which has been assumed. There are also other known effects, such as image force and tunnelling; these lie outside the scope of the present treatment, but give rise to effects which are too small to alter substantially the agreement between theory and experiment.

1.3 DETECTORS

The property of detection is potentially available from any non-linear resistance, i.e. a d.c. output may be derived which is proportional to the square of an applied alternating signal. To show this, consider a general I/V relationship represented as a Taylor expansion:

$$I_0 + \delta i = f(V_0) + f'(V_0)\delta v + \frac{f''(V_0)\delta v^2}{2!} + \frac{f'''(V_0)\delta v^3}{3!} + \cdots$$

where V_0, I_0 relate to the steady bias conditions and δv, δi refer to the instantaneous values of a small applied alternating signal. It is clear that for signals of sufficiently small amplitude the average increase in current is obtained from the third term of the expansion alone, the odd terms, of

Point-contact Crystal Diodes

course, making no resultant contribution. If δv is now used to represent the signal amplitude, the detected current (average increase) is

$$\delta i = \frac{f''(V_0)\delta v^2}{4} \qquad (1.6)$$

The signal power absorbed by the non-linear resistance may be written as

$$P = \frac{\delta v^2}{2} f'(V_0)$$

so that the current sensitivity is given by

$$\frac{\delta i}{P} \equiv \beta_0 = \frac{f''(V_0)}{2f'(V_0)} \qquad (1.7)$$

Now, β_0 is usually taken to refer to a detector operated with its output short-circuited and so represents the low-frequency short-circuit current sensitivity. For the exponential characteristic previously considered it follows that

$$\beta_0 = \frac{\alpha}{2}$$

The current sensitivity β appropriate to microwave frequencies is less than β_0 because of the degrading effect of the barrier capacitance. Taking the equivalent circuit as that of Fig. 1.2(a), to find β we assume that the signal is matched into the effective resistance appearing across the detector terminals, i.e. into $r + R'$, where R' is the series equivalent resistance of the parallel combination of R and C, given by

$$R' = \frac{R}{1 + \omega^2 C^2 R^2}$$

At microwave frequencies the current sensitivity is reduced in proportion to the ratio of signal voltage across R' to that applied to the detector terminals, i.e.

$$\beta = \frac{R'}{R' + r} \beta_0$$

$$= \frac{\beta_0}{1 + r/R + \omega^2 C^2 Rr}$$

$$\approx \frac{\beta_0}{1 + \omega^2 C^2 Rr} \quad \text{if } R \gg r \qquad (1.8)$$

Microwave Semiconductor Devices

If we include the voltage drop across the spreading resistance for both β_0 and β, a more correct result is obtained:

$$\beta = \frac{\alpha}{2\left(1 + \dfrac{r}{R}\right)^2} \frac{1}{1 + \dfrac{\omega^2 C^2 R^2 r}{R + r}} \tag{1.9}$$

1.3.1 Signal/Noise Ratio

For many purposes the current sensitivity is a very useful measure of detector quality, as for example in microwave measuring equipment, but with video receivers the question of noise becomes important also, and this is considered in terms of signal/noise ratio. Let R represent the video resistance of the detector, i.e. the sum of barrier resistance plus spreading resistance, measured at the operating bias point. For a given available microwave signal power P_S, the voltage output from a receiver of voltage gain A will be

$$V_S = A\beta P_S R$$

The available incremental noise power δP_N from any resistance in thermodynamical equilibrium with its surroundings, for an incremental frequency bandwidth δf, is given to a close approximation by

$$\delta P_N = kT\,\delta f$$

Assuming matched conditions, the noise voltage appearing at the receiver output may therefore be written

$$V_N = A[4kTB(R + R_A)]^{1/2}$$

where B is the effective noise bandwidth of the receiver and R_A is the equivalent input noise resistance, i.e. a ficticious resistance added to account for noise originating within the receiver itself.

The output signal/noise ratio is thus

$$\frac{V_S}{V_N} = \frac{\beta R}{(R + R_A)^{1/2}} \times \frac{P_S}{(4kTB)^{1/2}} \tag{1.10}$$

The factor $\beta R/(R + R_A)^{1/2}$ alone contains the detector parameters and is called the *figure of merit*, M.

Now, R_A is usually about 500 Ω, so that for detectors with video resistances of several kilohms it follows that $M \approx \beta\sqrt{R}$. Values for M are

10

typically 100 (watts)$^{-1/2}$ in the 3 cm band and fall to about 20 (watts)$^{-1/2}$ at a wavelength of 1 cm.

1.3.2 Tangential Senisitivity

The ability of a video detector to discriminate between small signals and a noise background is indicated by some criterion of limiting sensitivity such that the available signal power P_L at the video input is a fixed multiple of available input noise power, i.e.

$$P_L = \gamma P_N \qquad (1.11)$$

The particular limiting sensitivity represented by this equation depends on the value taken for γ. A commonly accepted criterion is that of *tangential sensitivity*, for which $\gamma = 6$ [11]. "Tangential" conditions are said to obtain, with an input signal in the form of a rectangular pulse, when the output voltage response, examined on an oscilloscope, has the apparent "edge" at the top of the standing noise in line with that at the bottom, lifted by the pulse. If we use the above value for γ and write P_T for the corresponding available microwave (tangential) power, eqn. (1.11) may be rewritten as

$$\frac{(\beta P_T)^2 R}{4} = 6kTB \frac{R + R_A}{R}$$

whence

$$P_T = \frac{1}{M}(24kTB)^{1/2}$$

The above expression makes clear the inverse relationship between tangential sensitivity and figure of merit. Values of $M = 100$ (watts)$^{-1/2}$, $B = 10^6$ Hz and $kT = 4 \times 10^{-21}$ J give $P_T = 3 \cdot 1 \times 10^{-9}$ W ≈ -55 dBm. Tangential sensitivity, for a 1 MHz bandwidth receiver, may therefore be written in terms of figure of merit as

$$P_T = \left(55 + 10 \log_{10} \frac{M}{100}\right) \text{dBm} \qquad (1.12)$$

1.3.3 Choice of Junction Parameters

The requirements for high sensitivity at a given microwave frequency may be expressed in terms of junction area and doping level. This amounts to minimizing the factor $C^2 Rr$, and for simplicity we will here ignore the possibility of an artificial depletion layer.

Microwave Semiconductor Devices

From eqn. (1.3) it follows that

$$C^2 \propto A^2 n$$

From eqn. (1.5), taking R as the slope resistance, we have, for small bias voltages which are insufficient to change the barrier potential appreciably,

$$R = \frac{1}{\alpha(I + I_0)} \tag{1.13}$$

With zero bias $I = 0$. Now, I_0 is clearly proportional to area, and from eqn. (1.4) it is also proportional to n. Thus

$$R \propto \frac{1}{nA}$$

For a non-epitaxial contact (see Fig. 1.3(a)),

$$r \propto \frac{\rho}{A^{1/2}}$$

where $\rho = 1/\mu e n$.

At the high doping levels appropriate to detector diodes, mobility, μ, is not strictly independent of n but is a decreasing function of doping level. In practice, detectors invariably employ p-type silicon, for which the hole mobility at the appropriate concentrations ($c.$ $10^{24}/\text{m}^3$) is approximately given by

$$\mu_n \propto \frac{1}{n^{1/4}}$$

which leads to

$$\gamma \propto \frac{1}{n^{3/4} A^{1/2}}$$

It follows, therefore, that the sensitivity reduction factor is related to n and A by the expression

$$C^2 Rr \propto \frac{A^{1/2}}{n^{3/4}}$$

Thus for high sensitivity we require small areas and high doping levels. Restrictions on area are set by mechanical stability and by the ability to

Point-contact Crystal Diodes

withstand electrical overload, whilst a limit on doping level arises as degeneracy in the semiconductor is approached and the I/V characteristic becomes less non-linear.

1.3.4 Biased Detectors

It is possible to make silicon crystals of improved tangential sensitivity by using forward bias. The reason for this becomes clear on considering the reduction of video resistance with bias indicated by eqn. (1.13) and

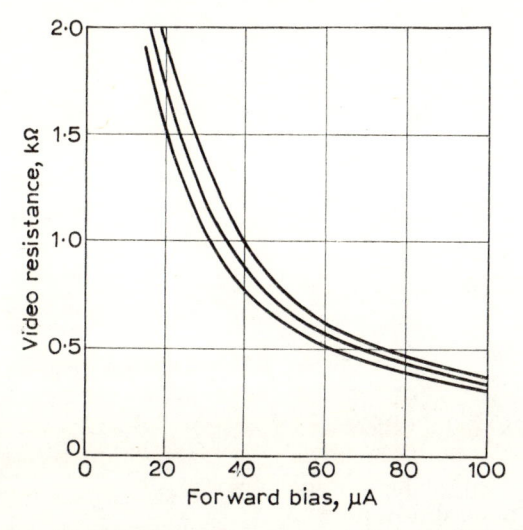

FIG 1.5 CV7180: variation of video resistance with forward bias: typical examples

the consequent effect of this on current sensitivity as expressed in eqn. (1.8). However, detectors intended to operate without bias do not in general show the expected improvement in tangential sensitivity, the reason being indicated below.

In order to get high doping levels, unbiased detectors use boron, which is soluble in silicon to the extent of more than 10^{26} atoms/m^3. Such detectors exhibit high values of I_0, but unless this parameter is small enough video resistance will not fall sufficiently rapidly with forward bias to offset the degrading effects of increasing contact capacitance and video noise.

Detectors which are improved by forward bias can be made from aluminium-doped silicon, the effect of bias on video resistance being shown in Fig. 1.5. The resulting improvement in tangential sensitivity for small

Microwave Semiconductor Devices

values of forward bias is shown in Fig. 1.6, the peak value being at least 3 dB better than for unbiased detectors.

Peak value of tangential sensitivity commonly occurs at around 20 μA forward bias, but a figure of 50 μA is usually specified. The reason for this is that, as the bias is progressively increased beyond the point corresponding to maximum sensitivity, the spread in r.f. admittance decreases

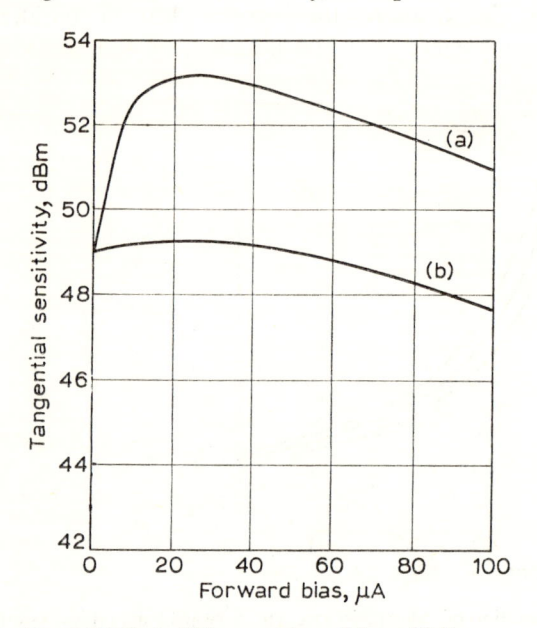

FIG 1.6 Effect of bias: on tangential sensitivity
$f = 9.375$ GHz
Matched conditions
(a) CV7180: biased detector
(b) CV2355: unbiased type

and a bias of 50 μA gives a fair compromise between sensitivity and spread. The results obtained for four specimens of a coaxial type of detector are shown in Fig. 1.7.

1.3.5 Broadband Design

It is usually necessary for a video detector to operate over a broad relative bandwidth, say 40%, which presents a special problem of mount design. Detectors are usually of coaxial construction, and it is found that the broadband performance can be accurately predicted by applying standard

Point-contact Crystal Diodes

transmission-line theory to the various coaxial sections within the crystal mount, making due allowance for the discontinuity capacitance between sections [12]. The aim is so to make use of the various electrical lengths and capacitances that the admittance associated with the point contact appears appropriately transformed at the input end of the mount. Here

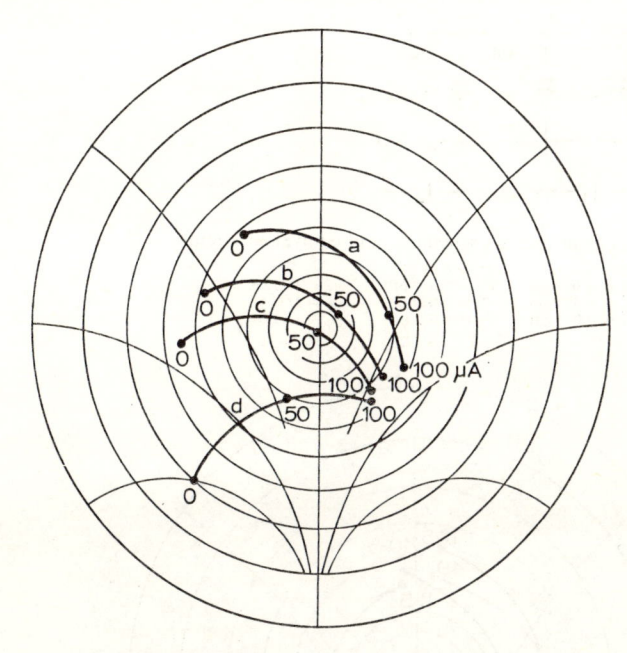

Fig 1.7 CV7180: variation of admittance with forward bias measurements on four samples in a holder designed to match mean admittance at 50 μA bias
$f = 9.375$ GHz

the admittance, normalized to that of the feeder and over the frequency bandwidth, should appear on a Smith chart as an approximately circular locus situated about its centre.

The greater the bandwidth which it is required to cover, the more will the locus spread out from the centre. The same is true for a given percentage bandwidth, as the frequency band is lowered, owing to the greater reactance of the contact capacitance. Figure 1.8 shows the derivation of the equivalent circuit for a coaxial mount for up to 40 GHz, and Fig. 1.9

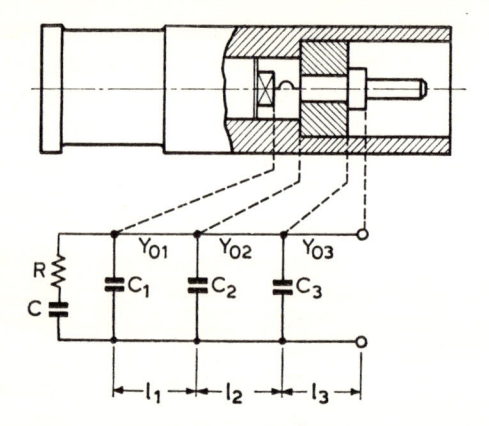

FIG 1.8 Coaxial mount: schematic and equivalent circuit

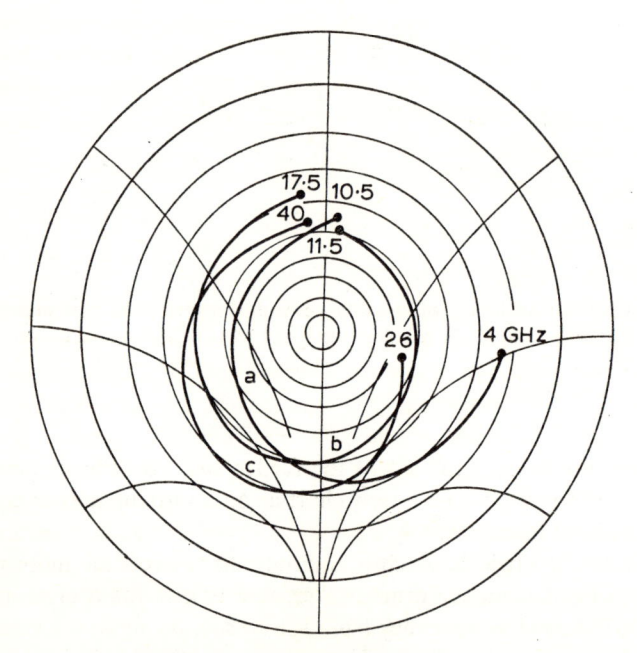

$R = 50\,\Omega$ $C = 0.05$ pF
$l_1 = 0.064$ cm $C_1 = 0.017$ pF $Y_{01} = 1/222$ S
$l_2 = 0.373$ cm $C_2 = 0.044$ pF $Y_{02} = 1/35$ S
$l_3 = 0.051$ cm $C_3 = 0.021$ pF $Y_{03} = 1/61$ S

FIG 1.9 Admittances of biased detectors
(a) CV7180   (b) CV7148   (c) CV7147

Point-contact Crystal Diodes

shows the admittance loci obtained for three biased detectors covering the frequency range 4–40 GHz.

1.3.6 Burnout

The ability to withstand electrical overload, or burnout, is important with all crystal diodes. With detectors the most important burnout effect is that due to prolonged exposure at modest overloads, the effect being a change in r.f. admittance which occurs before any impairment of sensitivity, the latter being indicative of heavy overloads. The admittance change is attributable to a decrease in contact capacitance and is associated with the formation of high-resistance layers [13]. The author has found that with 3 cm-band detectors an overload of 1 W peak power, 1 000 pulses/s, produced a mean shift in r.f. admittance in 5 min equal approximately to the diameter of a 0·8 v.s.w.r. circle on a Smith chart. Similar admittance changes were produced by 100 mW applied for one hour, and it appears that, at least over this power range, the deterioration is a function of the total energy received. For unbiased boron-doped detectors, the mean admittance shifts were much larger for the same energy and correspond approximately to the diameter of a 0·6 v.s.w.r. circle. Further, biased aluminium-doped detectors appeared stable after this exposure, whereas the boron type tended to drift, for several weeks afterwards, back towards their original admittance values.

1.4 MIXERS

1.4.1 Overall Noise Figure

The noise performance of a microwave heterodyne receiver is derived by considering it to be made up of two networks, the first representing the mixer and the second the i.f. amplifier together with its output circuit.

Let F and G denote respectively the noise figure and power gain, and suffixes 1 and 2 the first and second networks. Thus F_{1+2} and G_{1+2} will denote the overall noise figure and overall gain. In this context, *gain* is defined as the ratio of available power output to available power input, i.e. matched conditions are implied. By *noise figure* is meant the ratio by which a network is more noisy than an equivalent network generating only thermal noise; i.e. noise figure may be defined by the expression for the incremental output noise power

$$\delta N_0 = FGkT_0\, \delta f$$

where T_0 is a standard temperature, conveniently chosen as 290 K, and δf represents an incremental frequency band.

The available output noise power from the combined network over the incremental bandwidth δf is given by

$$\delta N_{0(1+2)} = F_{(1+2)}G_{(1+2)}kT\,\delta f$$
$$= F_{(1+2)}G_1G_2kT\,\delta f$$

since $G_{1+2} = G_1G_2$.

We may divide $\delta N_{0(1+2)}$ into the components arising from each stage, as follows. The component originating in the signal source and stage 1 is

$$N_0'_{(1+2)} = G_2\,\delta N_0$$
$$= F_1G_1G_2kT\,\delta f \tag{1.14}$$

The component originating in stage 2 is

$$\delta N_{0(1+2)} = F_2G_2kT\,\delta f - G_2kT\,\delta f \tag{1.15}$$

The term subtracted represents the noise from the output resistance of stage 1 which is included in eqn. (1.14).

Substitution in the equation

$$\delta N_{0(1+2)} = \delta N_0'_{(1+2)} + \delta N_0''_{(1+2)}$$

leads to

$$F_{(1+2)} = F_1 + \frac{F_2 - 1}{G_1}$$

If N_0 is the total noise output power which would be available if there were no noise sources within the network, the effective overall noise figure $F_{(1+2)}$ is given by

$$F'_{(1+2)} = \frac{N_0}{N_0'} = \frac{\int_0^\infty F_{(1+2)}G_1G_2 + G_2(F_2 - 1)\,df}{\int_0^\infty G_1G_2\,df} \tag{1.16}$$

In the present case of a heterodyne receiver the bandwidth of stage 2 is small compared with that of stage 1, so that G_1 and F_1 may be considered constant over the range of integration and eqn. (1.16) reduces to

$$F'_{(1+2)} = F_1 + \frac{F_2' - 1}{G_1} \tag{1.17}$$

Point-contact Crystal Diodes

In performing the integration indicated by eqn. (1.16) the factor G_2 is to be regarded as the gain of the network-meter combination, and this accounts for the prime on the right-hand side of eqn. (1.17).

For a crystal diode mixer, it is customary to write N_r for the product FG, the *noise ratio*. Instead of referring to the gain of a crystal mixer it is more appropriate to consider its reciprocal, L, the conversion loss. Thus, in eqn. (1.17),

$$N_r = F_1 G_1 \qquad L = \frac{1}{G_1}$$

which gives

$$F'_{(1+2)} = L(F_2' + N_r - 1)$$

or, with symbols more descriptive of the present situation, the effective overall noise figure of the receiver is given by

$$F'_{rec} = L(F'_{i.f.} + N_r - 1) \tag{1.18}$$

Another useful quantity is the *Y-factor*, which is defined as the ratio of available output noise power of an amplifier whose input terminals are loaded by the mixer to the same quantity when the amplifier is loaded by a dummy cartridge, containing an ohmic resistor. It may easily be shown that the Y-factor is then defined by the relationship

$$Y = \frac{N_r - 1}{F_{i.f.}} + 1 \tag{1.19}$$

which leads to a convenient way of determining N_r. It may further be shown that

$$F'_{rec} = L F'_{i.f.} Y \tag{1.20}$$

This expression has been widely utilized in equipments designed for the automatic measurement of overall noise figure, as indicated later.

1.4.2 Three-Stage Receiver

The above analysis can readily be extended to the case of several networks in cascade. It is particularly useful to do this for a receiver which has three stages, since the first one may be some form of r.f. amplifier, the second stage the mixer, and the third the i.f. amplifier/output-meter combination. The corresponding result for overall noise figure is

$$F'_{(1+2+3)} = \left(F_1 - \frac{1}{G_1}\right) + \frac{L}{G_1}(F_3' + N_r - 1) \tag{1.21}$$

which reduces to eqn. (1.18) when $F_1 = G_1 = 1$.

Eqn. (1.21) shows that, if the gain of the r.f. amplifier is large, the overall noise figure of the receiver is dominated by F_1. If G_1 exceeds 20 dB the other terms may usually be neglected.

With video receivers it may be shown that an initial stage of r.f. amplification (noise figure F_1, gain G_1 and bandwidth B_1) will increase the tangential sensitivity in proportion to G_1 up to a limit at which the product $F_1 G_1$ reaches a value given approximately by

$$F_1 G_1 = \frac{1}{\beta} \frac{4(F_2' + N_r - 1)}{kTB_1 R} \qquad (1.22)$$

where β and R are respectively the current sensitivity and video resistance of the detector [14]. Typical upper values for $F_1 G_1$ are of the order of 40 dB.

1.4.3 Frequency Conversion

We have noted that the two attributes of crystal diode mixers which determine receiver overall noise figure are conversion loss and noise ratio. Since in modern mixers the latter attribute can be made close to unity, it is the conversion loss which is of prime interest and this we now consider. A detailed treatment of the problem of frequency conversion is given by Torrey and Whitmer [9], and we shall follow their treatment in outline.

We consider a non-linear resistance to whose terminals are applied two frequencies, one ω at high level and the other α at low level, the former representing the local oscillator and the latter the signal. The function of a mixer is to convert the signal frequency to the intermediate frequency $\beta = \alpha - \omega$ as a result of beating α with ω. Many other frequencies are created, however, in particular the sum frequency $\alpha + \omega$, which is generated with almost equal efficiency, but those of principal importance are as follows:

At low level $\begin{cases} \alpha \equiv \omega + \beta & \text{(signal)} \\ \beta & \text{(intermediate frequency)} \\ \gamma \equiv \omega - \beta & \text{(image)} \\ n\omega \pm \beta & \text{(harmonic sidebands)} \end{cases}$

At high level $\begin{cases} \omega & \text{(local oscillator)} \\ n\omega & \text{(harmonics of the local oscillator)} \end{cases}$

As the signal, image and intermediate frequencies are at very low levels, the relations among their voltages and currents are linear, and the mixer may therefore be regarded as a linear network with separate terminals at each of these three frequencies. Physically, of course, the signal and image terminals are identical, but conceptually they may be regarded as distinct, since they can be separately loaded by means of sharply tuned circuits.

We may write a set of linear relations between the currents and voltages at the three frequencies as

$$I \begin{bmatrix} \alpha \\ \beta \\ \gamma^* \end{bmatrix} = Y \begin{bmatrix} V_\alpha \\ V_\beta \\ V_\gamma^* \end{bmatrix}$$

where

$$Y \equiv \begin{bmatrix} y_{\alpha\alpha} & y_{\alpha\beta} & y_{\alpha\gamma} \\ y_{\beta\alpha} & y_{\beta\beta} & y_{\beta\gamma} \\ y_{\gamma\alpha} & y_{\gamma\beta} & y_{\gamma\gamma} \end{bmatrix}$$

is the admittance matrix of the mixer. It is shown by Torrey and Whitmer that the image current and voltage must appear as complex conjugates in these relations. In the general form the admittance matrix consists of nine complex quantities, each consisting of two components, making 18 parameters in all, but in practical cases this number can be considerably reduced so that the matrix takes the form

$$Y = \begin{bmatrix} g_{\alpha\alpha} & g_{\alpha\beta} & g_{\alpha\gamma} \\ g_{\beta\alpha} & g_{\beta\beta} & g_{\beta\alpha} \\ g_{\alpha\gamma} & g_{\alpha\beta} & g_{\alpha\alpha} \end{bmatrix}$$

We now consider the I/V characteristic of the mixer to be given by

$$I = f(V)$$

and assume the current and voltage to be composed of the following parts:

(a) I_0 V_0 D.C.
(b) ΣI_n $V e^{jn\omega t} \Sigma V_n e^{jn\omega t}$ Local oscillator and harmonics
(c) Re $\Sigma I_\mu e^{jb\mu t}$ Re $\Sigma V_\mu e^{jb\mu t}$ Signal, image and harmonic sidebands

In (b) the summations extend over all integral values of n from $-\infty$ to $+\infty$, excluding 0. In (c) b_μ includes the frequencies $\omega + \beta$, $\omega - \beta$, β and $n \pm \beta$. It is assumed that the terms in (c) are very small compared with those in (a) and (b) and can be considered to be variations of them. The general variational equation is then

$$\delta i = \frac{dI}{dV} \delta v$$

where δi and δv are identified with the current and voltage terms in (c). For dI/dV we may use a Fourier expansion:

$$\frac{dI}{dV} = \sum_{n=-\infty}^{n=+\infty} y_n e^{jn\omega t} \qquad (1.23)$$

It may then be shown that the admittance matrix is given by

$$Y = \begin{bmatrix} y_0 & y_1 & y_2 \\ y_1^* & y_0 & y_1 \\ y_2^* & y_1^* & y_0 \end{bmatrix}$$

If we assume the exponential I/V characteristic,

$$I = I_0(e^{\alpha V} - 1)$$

then

$$\frac{dI}{dV} = \alpha I_0 \, e^{\alpha V}$$

Neglecting harmonics of the local oscillator voltage and writing

$$v = V_0 + V_1 \cos \omega t$$

leads to the Fourier expansion

$$\frac{dI}{dV} = \alpha I_0 \, e^{\alpha V_0} = \sum_{n=-\infty}^{n=+\infty} I_n(\alpha V_1) \, e^{jn\omega t} \qquad (1.24)$$

where $I_n(x)$ is the modified Bessel function $j^{-n} J_n(jx)$. Comparing this with eqn. (1.21), we see that

$$y_0 = \alpha I_0 \, e^{\alpha V_0} I_0(\alpha V_1) = g_{\alpha\alpha} = g_{\beta\beta}$$
$$y_1 = y_1^* = \alpha I_0 \, e^{\alpha V_0} I_1(\alpha V_1) = g_{\alpha\beta} = g_{\beta\alpha}$$
$$y_2 = y_2^* = \alpha I_0 \, e^{\alpha V_0} I_2(\alpha V_1) = g_{\alpha\gamma}$$

The conversion loss is defined by

$$L = \frac{\text{Available power at signal frequency}}{\text{Available power at intermediate frequency}}$$

and the various conditions under which it is customary to specify L are as follows:

- L_0 Mixer matched to local oscillator, same termination at signal and image frequencies
- L_1 Mixer matched to signal, image short-circuited
- L_2 Minimum loss with the same termination at signal and image frequencies
- L_3 Mixer matched to signal, image open-circuited

It may be shown for an admittance matrix such as that indicated above, which is entirely real,

$$L_0 = 2k \left(\frac{1}{\varepsilon_1} + \frac{1}{\varepsilon_3} - 1 \right)$$

$$L_1 = k \left[\frac{1 + \sqrt{(1 - \varepsilon_1)}}{1 - \sqrt{(1 - \varepsilon_1)}} \right]$$

$$L_2 = 2k \left\{ \frac{1 + \sqrt{[(1 - \varepsilon_1)(1 - \varepsilon_3)]}}{1 - \sqrt{[(1 - \varepsilon_1)(1 - \varepsilon_3)]}} \right\}$$

$$L_3 = k \left[\frac{1 + \sqrt{(1 - \varepsilon_3)}}{1 - \sqrt{(1 - \varepsilon_3)}} \right]$$

where

$$k = \frac{g_{\alpha\beta}}{g_{\beta\alpha}}$$

$$\varepsilon_1 = \frac{g_{\alpha\beta} g_{\beta\alpha}}{g_{\alpha\alpha} g_{\beta\beta}}$$

$$\varepsilon_3 = \frac{\varepsilon_1}{1 - \varepsilon_1} \frac{1 - \theta}{1 + \theta}$$

with $\theta = g_{\alpha\gamma}/g_{\alpha\alpha}$.

For the exponential I/V characteristic it follows that

$$\varepsilon_1 = \frac{I_1^2}{I_0^2} \qquad \varepsilon_2 = \frac{I_1^2}{I_0^2 - I_1^2} \frac{I_0 - I_2}{I_0 + I_2}$$

The most commonly quoted value of conversion loss, L_0, is then given by

$$L_0 = \frac{4I_0(I_0^2 - I_1^2)}{I_1^2(I_0 - I_2)} \tag{1.25}$$

Microwave Semiconductor Devices

All the values of L are decreasing functions of local oscillator drive. It is found that L_0 and L_2 are nearly equal and their lowest possible value is 3 dB, whereas L_1 and L_3 can in principle come down to zero. In practice, of course, a crystal mixer is not simply a non-linear resistance, as indicated by the equivalent circuit of Fig. 1.2(b), and the effect of the parasitics is to prevent the theoretical figures for conversion loss being realized in practice. The best results which have been reported for a point-contact structure

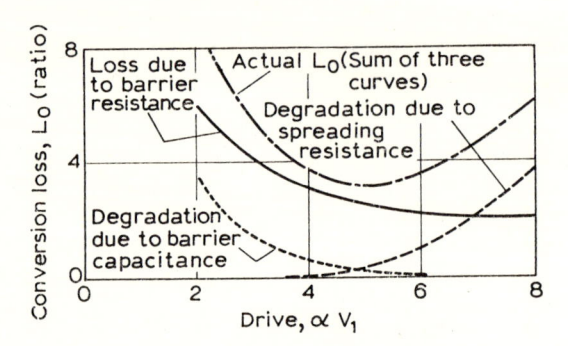

FIG 1.10 Variation of conversion loss with drive
(T.H.B. Baker—private communication)

relate to germanium with titanium whiskers. These are 4·2, 4·7 and 2·2 dB respectively for L_0, L_1 and L_2, with a signal frequency of 10 GHz [4].

When we consider the effect of spreading resistance it is found that conversion loss still falls with local oscillator drive initially, but as the average value of barrier resistance is reduced to a level which approaches that of the spreading resistance, more power is dissipated in the latter and the conversion loss increases again. Figure 1.10 shows this effect in relation to L_0. This information, together with much of the remainder of the present section, has been drawn from unpublished extensions to the treatment of Torrey and Whitmer by T. H. B. Baker.

The influence of barrier capacitance is most pronounced at low local oscillator drive since it then most effectively shunts the barrier whose resistance is at its greatest. The ratio of power which is available into the average barrier resistance to the total input power is found to be

$$\frac{P_b}{P_a} = \frac{1}{1 + r/R + \omega^2 C^2 Rr}$$

Point-contact Crystal Diodes

This expression is, of course, the same as that previously encountered in connection with the current sensitivity of detector crystals. If R is made adjustable (by local oscillator drive level or the use of d.c. bias) it is readily shown that maximum power is delivered to the barrier when

$$R = \frac{1}{\omega C}$$

whence

$$\left(\frac{P_b}{P_a}\right)_{max} = \frac{1}{1 + 2\omega Cr}$$

The minimum attainable loss at high frequency, allowing for barrier capacitance, is therefore

$$L_c = L_0(1 + 2\omega Cr) \tag{1.26}$$

This result is somewhat optimistic as it neglects second-order effects resulting from interaction terms. When these are considered, the salient terms in the admittance matrix are modified, to a good approximation, as follows:

$$g_{\alpha\alpha} = g_0\left[1 + rg_0\left(2 - \frac{g_1^2 + g_2^2}{g_0^2}\right) + \mu^2\left(1 + \frac{1}{rg_0}\right)\right]$$

$$g_{\beta\beta} = g_0\left[1 + rg_0\left(2 - 2\frac{g_1^2}{g_0^2}\right) + \mu^2\right]$$

$$g_{\alpha\beta} = g_1\left[1 + rg_0\left(1 - \frac{g_2}{g_0}\right)\right]$$

$$g_{\alpha\gamma} = g_2[1 + rg_0(1 - g_1^2/g_0g_2)]$$

where $\mu = \omega Cr$, and g_0, g_1, g_2 are the first three terms in the expansion of dI/dV.

The condition $R = 1/\omega C$ for minimum L_0 determines the drive αV_1 and mean (or rectified) current I_{mean}. The mean barrier conductance is of course g_0, given in terms of the modified zero-order Bessel function of argument αV_1 by

$$g_0 = \alpha I_0 e^{\alpha V_0} I_0(\alpha V_1)$$
$$= \alpha I_0 I_0(\alpha V_1)$$

in the absence of d.c. bias. For an exponential characteristic we also have

$$g_{mean} = \alpha I_{mean}$$

Thus

$$\frac{1}{R} = g_0 = \alpha I_{\text{mean}} = \alpha I_0 I_0(\alpha V_1) = \omega C$$

whence

$$I_0(\alpha V_1) = \frac{\omega C}{\alpha I_0} \qquad (1.27)$$

and

$$I_{\text{mean}} = \frac{\omega C}{\alpha} \qquad (1.28)$$

From eqn. (1.26) it follows that, for minimum conversion loss, the product Cr should be minimized. Writing N for the charge density en, eqn. (1.3) shows that

$$C \propto A\epsilon^{1/2} N^{1/2}$$

and, from the expressions following eqn. (1.13),

$$r \propto \frac{1}{\mu N A^{1/2}}$$

so that

$$Cr \propto \frac{A^{1/2} \epsilon^{1/2}}{\mu N^{1/2}}$$

The contact area A is usually determined by considerations of mechanical stability and resistance to burnout and not, as are the remaining factors, by the particular semiconductor material. If we ignore the weak dependence of mobility on carrier concentration, a figure of merit for the material may be defined as

$$M = \frac{N^{1/2} \mu}{\epsilon^{1/2}} \qquad (1.29)$$

which should be maximized for minimum loss. The optimum value of N occurs at about [15]

2×10^{17} carriers/cm² for gallium arsenide
1×10^{18} carriers/cm² for germanium
5×10^{18} carriers/cm² for silicon

On the basis of this figure of merit we find that the rating of the materials is in the order in which they are quoted above (see Fig. 1.11).

Point-contact Crystal Diodes

We have considered the question of the degradation of intrinsic loss by spreading resistance and barrier capacitance, but there is also the possibility of reducing the intrinsic loss itself. Since this increases with local oscillator drive we require the optimum drive to be as high as possible, assuming there are no other restrictions on drive level. This means that the factor $\omega C/\alpha I_0$

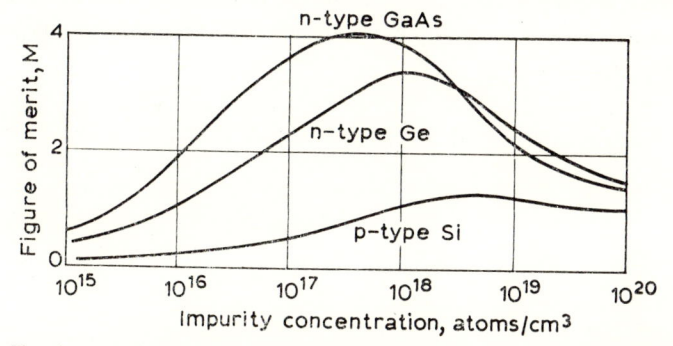

Fig 1.11 Variation of figure of merit with doping level for different semiconductors
(*T.H.B. Baker—private communication*)

must be large. In the numerator ω is fixed, and if C is increased then r must be decreased in the same ratio for the degradation factor not to increase. In the denominator, we note that a low α is beneficial rather than otherwise, except that a higher driving voltage is then required. The remaining parameter I_0 must be as small as possible.

Figure 1.12 shows the optimum value of L_0 as a function of αI_0 for different C and r. It is of interest to note that of the two curves having the same Cr product the one with the higher value of C gives the lower loss.

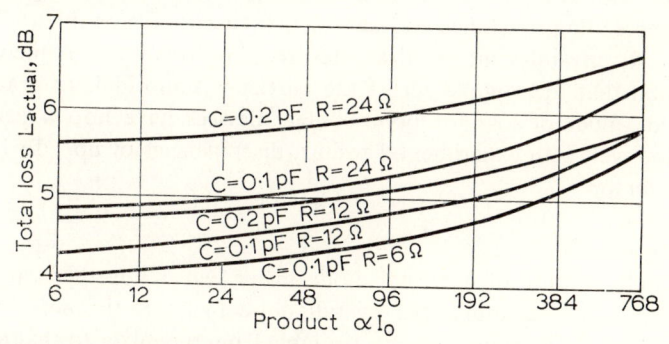

Fig 1.12 Optimum value of conversion loss L_0 versus αI_0
(*T.H.B. Baker—private communication*)

27

Since I_0 is directly proportional to $e^{-e\phi_0/kT}$ a high barrier would appear to be advantageous. The barrier height found with gallium arsenide (0·6–0·8 eV) is about twice that for silicon and germanium mixers, so that gallium arsenide is also indicated as the best material on this basis.

The application of local-oscillator drive to a mixer crystal produces a time-varying conductance which may be expanded as a Fourier series of sine or cosine terms. For example, taking a series of cosines, we may write the conductance as

$$g(t) = \frac{g_0}{2} + g_1 \cos \omega t + g_2 \cos 2\omega t + \ldots$$

where

$$g_n = \frac{2}{T} \int_{-T/2}^{T/2} g(t) \cos n\omega t \, dt$$

and $T = 2\pi/\omega$.

Comparison with the earlier results, in which the expansion for conductance was given in terms of a series of exponential terms, enables the various conversion losses to be written in terms of g_0, g_1, g_2. For instance, in the case of a mixer matched to the local oscillator, with the same termination at signal and image frequencies, we find that

$$L_0 = \frac{4g_0(g_0^2 - g_1^2)}{g_1^2(g_0 - g_2)} \tag{1.30}$$

Thus it is possible to calculate figures for the intrinsic conversion losses associated with any arbitrary I/V characteristics. In this way, by lumping together spreading resistance and barrier resistance, it is possible to overcome a difficulty inherent in all of the preceding treatment, namely the assumption that the voltage across the barrier is sinusoidal. In practice, calculated values of loss for point-contact devices have not in general agreed too well with experimental results, discrepancies of up 1 dB being not uncommon.

1.4.4 Noise Ratio

The second attribute of a crystal diode mixer, which together with the conversion loss, determines the overall noise figure is the *noise ratio*, defined as the ratio of noise power available from the mixer to that available, $kT_0 \delta f$, from an equivalent resistor at the reference temperature

$T_0 = 290$ K, this resistor having a value equal to the i.f. resistance of the mixer. In older literature the term "noise temperature ratio" is often used, but the association of excess noise with temperature in this particular instance is somewhat unnecessary and confusing.

There may be considered to be three sources of noise within a crystal diode mixer. The ohmic spreading resistance produces thermal noise and the barrier produces shot noise and also flicker noise. The origin of the latter is still not completely understood. The situation was probably summed up correctly by Nicoll [16] when he concluded that flicker noise usually dominates the noise behaviour at frequencies as high as 45 MHz. By selecting crystals whose flicker noise was low, however, he established good agreement between noise measured at 45 MHz for low bias current and the calculated values of shot noise.

Thermal or Johnson Noise

The random voltage fluctuations between the terminals of any resistance in thermodynamic equilibrium may conveniently be described in terms of current generators (per unit bandwidth) feeding the appropriate resistance. For thermal noise a short-circuited noise generator feeding a resistance R is described by a mean square current fluctuation

$$\overline{I^2} = \frac{4kT}{R} \text{ amperes}^2 \text{ per unit bandwidth}$$

(The condition of thermodynamic equilibrium implies that no power is received from any external energy source such as a battery.)

Shot Noise

Because of the discrete nature of charge carriers the value of a current is the statistical average of a number of elementary events and its instantaneous value therefore fluctuates slightly. For a current I in which there is no correlation between the motion of the individual charges of magnitude e and in which the time of transit of the elementary charge through the system is negligibly short, it may be shown that the mean square current fluctuation is

$$\overline{I^2} = 2eI$$

It can be shown from kinetic theory that thermal noise has the same nature and origin as shot noise.

Flicker Noise

Measurements of noise on widely varying items such as thin metallic films, thermionic valves with oxide coatings, carbon resistors and copper-oxide, silicon and germanium rectifiers have shown that the noise measured at low frequencies, when current is flowing, is much in excess of shot or thermal noise. All show the excess noise to be approximately of the form

$$\overline{I^2} \propto \frac{I^m}{f^n}$$

where I is the mean current, f the frequency, and m and n are around 2 and 1 respectively but may be slowly-varying functions of frequency. It is convenient to describe this type of contribution generically as flicker noise.

It appears to be generally accepted that flicker noise in point-contact diodes is due to modulations of the Schottky barrier potential at emitting patches by random fluctuations in the concentration of mobile adsorbed ions. However, it is necessary to postulate a spread in the values for the diffusion activation energy to account for both the f^{-1} variation and the experimentally found lack of marked variation with temperature.

Mechanical instability can give rise to a form of noise in a badly made contact or if splitting of the whisker material occurs, a situation which must certainly be guarded against with tungsten. There seems, however, to be little evidence to support a suggestion made by Schiff [17] that thermal instability of the several spots of contact between the whisker and semiconductor gives rise to noise.

Discussion of Noise

A detailed treatment of shot noise in a rectifying contact was given by Weisskopf [18]. He assumed that the current/voltage relationship is of the ideal exponential form of eqn. (1.5), in which we noted that the first term represents current in the forward direction and the second indicates that in the opposite direction.

The shot noises associated with these two distinct currents over the barrier must be added together as they are uncorrelated, giving for the total shot noise, with neglect of transit time and space-charge smoothing effects,

$$\overline{I^2} = 2eI_0 \, (\mathrm{e}^{eV/kT} + 1)$$

The noise ratio is the quotient of available shot-noise power by thermal noise per unit bandwidth, and is given by

$$N_r = \frac{\overline{I^2}R}{4kT}$$

where R is the slope resistance of the mixer at its operating point. R is clearly given by

$$\frac{1}{R} = \frac{dI}{dV}$$

but, in order to derive an expression for R which allows for the possibility of a variation of barrier potential with voltage, we must refer to eqn. (1.4) and replace I_0 in eqn. (1.5) by $I_0' \, e^{-e\phi_0/kT}$. This leads to

$$\frac{dI}{dV} = \frac{eI_0'}{kT} e^{-e\phi_0/kT} \left[e^{eV/kT} - \frac{d\phi}{dV}(e^{eV/kT} - 1) \right]$$

and hence

$$N_r = \tfrac{1}{2} \frac{1 + e^{-eV/kT}}{1 - \dfrac{d\phi}{dV}(1 - e^{-eV/kT})} \qquad (1.31)$$

To indicate the manner in which shot noise varies with bias we will consider three limiting cases, namely zero bias, large positive bias and large negative bias.

Since $e^{-eV/Tk}$ is unity at zero bias we therefore have $N_r = 1$. This clearly shows that thermal noise may be regarded as a limiting form of shot noise, when the mixer is in thermodynamic equilibrium.

With large positive bias the factor $e^{-eV/kT}$ tends to zero, and the noise ratio therefore tends to

$$N_r = \frac{1}{2(1 - d\phi/dV)} \qquad (1.32)$$

For the simple Schottky barrier $d\phi/dV$ would be zero and the noise ratio would tend to a value of one-half for positive bias. For most silicon rectifiers, however, it is found that $d\phi/dV$ is about one-half and the noise ratio is therefore about unity.

Microwave Semiconductor Devices

With large negative bias the terms in $e^{-eV/kT}$ have a dominant effect and the noise ratio tends to

$$N_r = \frac{1}{2(d\phi/dV)} \tag{1.33}$$

For silicon mixers under negative bias, $d\phi/dV$ is of the order of 0·1 and the noise ratio is therefore several times unity. They thus have a much higher noise ratio when biased in the reverse direction than when biased forward.

Microwave Noise

Noise is known to be produced well into the microwave spectrum when large reverse bias currents are passed through point-contact diodes, and for this reason they were formerly used as noise generators. They required frequent calibration for this purpose, however, and have long since been replaced by gas discharge tubes. The noise ratio at microwave frequencies is substantially unity for positive bias and increases linearly with negative bias current. It may be shown that this behaviour is consistent with the shot noise mechanism. However, it has also been shown that in certain conditions it may be possible to observe flicker noise at frequencies as high as 10^{10} Hz.

Noise under Local-Oscillator Excitation

The general problem of noise with local-oscillator excitation has not been analysed rigorously, but some visualization is possible from the foregoing discussion. In particular, it follows that low-noise mixers are required to have high reverse impedance (approximately $1 \text{M}\Omega$ at -1V), and this is much easier to achieve with silicon than with germanium, thus giving silicon a distinct advantage in terms of noise ratio, although the reverse is true for conversion loss.

In the past, point-contact mixers tended to become more noisy as the operating wavelength for which they were designed increased. This is consistent with a flicker noise effect, which becomes more pronounced as the area of contact is made smaller in an atteopt to maintain sensitivity. For a given current a smaller contact area implies on average a greater flicker noise, whereas shot noise is unaltered by changes in area. This can easily be seen by considering two identical contacts in parallel, each carrying a current I, compared with a single contact carrying the same total current $2I$. Since the mean-square noise current depends on the square

Point-contact Crystal Diodes

of the bias current, the total noise will be proportional to $2I^2$ in the case of the two contacts in parallel, compared with a value proportional to $4I^2$ for the single contact.

The trend towards higher noise with smaller areas is not so pronounced with modern point-contact devices, in which the surface preparation has been optimized to minimize the flicker effect, since the noise ratio for shot noise depends only on the total current and not on the area. It is invariably found that when crystals are excessively noisy this is due to a flicker effect, which is best measured at audio frequencies. Figure 1.13 shows good

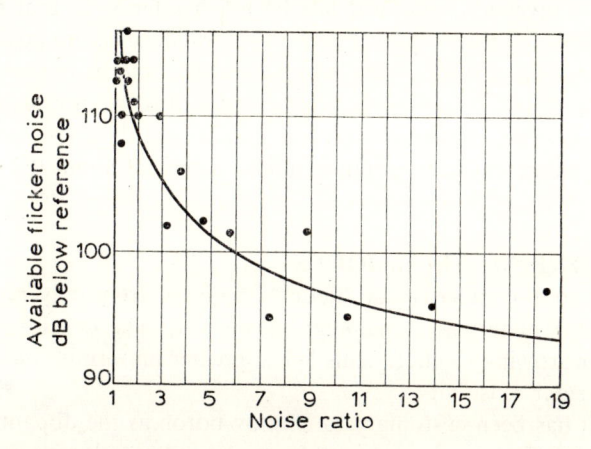

Fig 1.13 Available flicker noise at 1 kHz against noise ratio at 45 MHz (3 mA local-oscillator drive)

correlation obtained by the author for excessively noisy silicon diodes between flicker noise measurement at 1 kHz and the noise ratio obtained at 45 MHz under 3 cm local oscillator excitation.

1.4.5 Mixers for C.W. Radars

In recent years radars have been introduced which depend on the Doppler principle, and with these any noise contribution at audio frequencies is most important [19, 20]. Flicker noise in the mixer, which can usually be made negligibly small for pulse radar applications on account of the relatively high intermediate frequency, is therefore a serious consideration and has led to the development of mixers with minimum flicker effect. The best results have been obtained with single-crystal silicon heavily doped with aluminium, in which the reverse current is minimized by surface etching and subsequent heat treatment. However, it should be added that recent advances in germanium point-contact technology are challenging the superiority of silicon in these applications.

1.4.6 Harmonic Mixers

The frequencies generated by the mixing of a signal with a harmonic of the local oscillator are normally regarded as spurious. They can, however, be exploited to extend the frequency range if the mixer is optimized for such harmonic operation [21, 22]. It has been reported that low-order harmonic mixer conversion loss need not be much inferior to that of a fundamental mixer, and that improvements of several decibels are possible under conditions of reverse bias. This behaviour appears to be in contrast to that of fundamental mixers, where conversion loss is not usually improved with d.c. bias, except for certain germanium diodes in which optimum performance has been obtained under a bias of about 0·2 V in the forward direction.

1.5 MANUFACTURING TECHNIQUES

For many years point-contact crystal diodes were made from polycrystalline semiconductor materials, principally silicon [23]. The development of techniques for growing single crystals led to greater uniformity but not necessarily to greater sensitivity.

With silicon it has been customary to employ boron as the dopant for unbiased detectors and aluminium for biased detectors and mixers, the resistivity being of the order of 0·01 Ω-cm. Heat treatment for several hours at around 900°C is employed to create an artificial depletion layer through the out-diffusion of impurity atoms. This is found to be particularly advantageous for aluminium-doped material in which the principal effect is to reduce the reverse leakage, thereby improving the noise performance [1]. With germanium, n-type dopants, particularly phosphorus, are employed with resistivities of the order of 0·002–0·006 Ω-cm, and it is beneficial to polish the surface mechanically [4]. The usual method with gallium arsenide is to dope purified material by re-growing crystals in an atmosphere of arsenic which also contains donor impurities, e.g. sulphur, selenium or tellurium. For such material resistivities between 0·06 and 0·002 Ω-cm have been investigated [6].

Tungsten and molybdenum-tungsten alloy whiskers are normally used with silicon, and the forming process after contact is established consists in applying a few sharp taps to the diode cartridge. Titanium whiskers are

Point-contact Crystal Diodes

used with germanium and phosphor-bronze with gallium arsenide, the forming with these being effected by applying one or more electrical pulses. During the forming process it is usual to monitor the forward and reverse current as well as either the contact capacitance or the r.f. admittance.

1.6 MEASURING TECHNIQUES

1.6.1 Detectors

The most basic measurement made on unbiased detectors is that of current sensitivity. It would appear that the simplest way to do this is to measure the rectified current into a low-impedance microammeter when a microwave signal of the order of a microwatt is applied. However, for reasonable

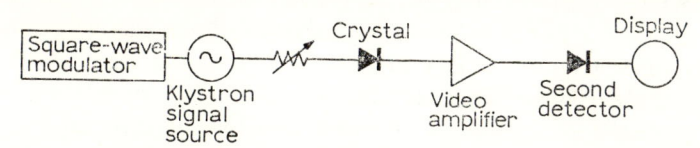

Fig 1.14 Tangential sensitivity equipment

sensitivity, a level of a few microwatts is desirable and the resulting rectified current of a few microamperes will lower the video resistance sufficiently to give an optimistic value of current sensitivity. A more accurate value is obtained from the open-circuit voltage produced under similar drive conditions by means of a voltage divider. The rectified current under short-circuit conditions is then given by this voltage acting across the video resistance.

Video resistance is conveniently measured by using an audio-frequency source giving a constant current output of the order of a microampere. The voltage developed across the video resistance of the crystal by this source is then compared with that produced across calibrating resistances, using a high-sensitivity tuned audio amplifier. If this has a linear output meter it can be made direct reading in resistance.

With biased detectors a measurement of tangential sensitivity is required, and this is usually obtained using the type of equipment illustrated schematically in Fig. 1.14. If the video amplifier has a bandwidth of 1 MHz it is convenient to modulate the klystron at a frequency of about 50 kHz using a square wave. A calibrated attenuator is adjusted to give a visual picture on the output display tube corresponding to that defining "tangential" conditions. After the equipment has been calibrated with any appropriate power-measuring device, the attenuator setting may be made direct reading in terms of decibels relative to 1 mW. It is possible to improve the accuracy of measurement by replacing the display tube with a meter appropriately calibrated, although it is then desirable to use a cathode-ray tube for monitoring.

A further r.f. measurement applicable to all mixers and detectors is that of admittance or voltage standing-wave ratio (v.s.w.r.) in a specified mount. To this end, general-purpose instruments may be used, either a standing-wave indicator or an automatic admittance display equipment.

Detectors used in broad-band video receivers are subject to a form of electrical overload or *burnout* due to exposure to pulses of microwave power of, say, microsecond duration at 1 000 pulses/s. As a result the detectors are liable to undergo permanent changes of r.f. admittance without any marked deterioration in sensitivity for moderate overloads, and to suffer large reductions in r.f. sensitivity under heavy overloads. The existence of these effects requires the provision of suitable test equipment to assess the resistance of video detectors to specified overload conditions. It is customary to use a pulse-modulated magnetron for this purpose, feeding the bulk of the output into a high-power matched termination and diverting an appropriate fraction into the diode under test. Measurements of sensitivity and/or r.f. admittance are made before and after exposure. Reverse current at, say, -1 V will give an indication of severe burnout, when the leakage shows a marked increase.

1.6.2 Mixers

The most useful single measurement on a mixer is that of overall noise figure, and the usual bench set-up for doing this is indicated in Fig. 1.15. The mixer under test is excited at the normal local-oscillator level, say 500 μW, a filter with a Q-factor exceeding 1 000 being used to remove oscillator noise.

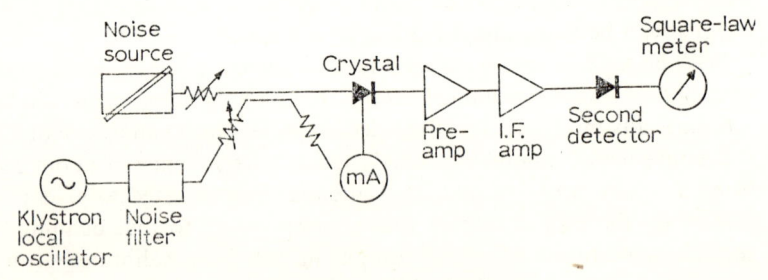

Fig 1.15 Equipment for measuring overall noise figure

Point-contact Crystal Diodes

A signal is added to this excitation which is derived from a standard noise tube whose output is some 15 dB above thermal noise level. This results in a wide spectrum of frequencies being produced by mixer action, and those which cover the pass-band of the i.f. amplifier are transmitted by it to give a response on the output meter.

In practice the i.f. system is divided into a pre-amplifier of low noise figure and a post-amplifier which has higher gain but whose noise figure is less important. Between the two is connected an i.f. pushbutton attenuator. By means of an r.f. attenuator in series with the noise source the noise level can be set so that, with the source switched on and 3 dB added on the pushbutton attenuator, the same output indication is obtained as with the source off and the 3 dB removed. The noise power incident at the mixer which satisfies this condition (expressed as a multiple of thermal power, kTB) is then equal to the overall noise figure as previously defined. Correct matching is, of course, necessary between each of the various components.

The noise power at the mixer is obtained by subtracting from the known output of the noise source the total insertion loss up to the mixer together with the added attenuation. Once the system has been calibrated the attenuator may be made to read overall noise factor directly. It must be remembered that with a broad-band mixer the microwave noise at frequencies above and below the local oscillator frequency will be equally effective in producing an i.f. response so that the effective noise power from the source is doubled. There is a further point in relation to the calibration of the attenuator when used in this type of application, as discussed below.

Consider a matched termination in waveguide at a temperature T_0 and therefore radiating an available power $kT_0 \delta f$ per unit bandwidth. When this is viewed through an attenuator set to attenuate by a factor A and at the same temperature T_0, the total available power from the attenuator will still be $kT_0 \delta f$, but of this a fraction $1/A$ must be attributed to the matched termination, leaving the power attributable to the attenuator as $kT_0 \delta f (1 - 1/A)$. On replacing the matched termination by a noise source of effective temperature T_e the total available noise power will now consist of a contribution $kT_e \delta f / A$ from the noise source together with an amount $kT_0 \delta f (1 - 1/A)$ from the attenuator. The available power, P, is thus given by

$$P = \frac{k \delta f}{A} [T_e + (A - 1)T_0] \qquad (1.34)$$

In practice, correction for the above effect can usually be neglected.

Microwave Semiconductor Devices

Conversion loss may conveniently be obtained by making a fixed change in the i.f. attenuator setting and noting the change in r.f. attenuator reading to maintain the output meter reading at its original value. Assuming matched conditions and bearing in mind the foregoing comments on effective attenuation, conversion loss is given as the ratio of the changes in r.f. and i.f. attenuation respectively.

It is usual to build into the pre-amplifier an i.f. noise diode by means of which the amplifier noise figure can be measured. It can easily be shown from the formula for shot noise that the noise ratio of a resistor R connected to a noise diode is

$$N_r = 1 + 20IR \tag{1.35}$$

where I is the diode current. This resistor may be mounted separately or introduced via a dummy cartridge which is inserted into the crystal mount. The noise figure of the amplifier is given by the value of $20IR$ at the diode current which just doubles the noise output corresponding to zero current or alternatively gives the same output reading when 3 dB of i.f. attenuation is introduced.

The noise ratio of the mixer may be found from eqn. (1.18), once the other parameters have been determined. Alternatively it may be obtained as an independent measurement from the Y-factor (eqn. (1.19)), the calibration again employing a resistor made noisy by an i.f. noise diode. However, when this direct procedure is adopted a separate amplifier is used with an input coupling circuit, such as that attributed to Roberts [24], which makes the Y-factor independent of crystal impedance to a first approximation and also serves as an impedance transformer matching the mixer to the amplifier.

With mixers, a measurement of i.f. impedance is also required and this is conveniently achieved using a method similar to that described for the video resistance of detectors. Instead of an audio signal, however, a small i.f. signal from a constant-current source is used. This is fed into the i.f. terminals of the mixer, which is operated at its normal impedance level by the application of local-oscillator drive. The i.f. voltage developed across the impedance of the mixer, which is predominantly resistive at such frequencies, is used as an indication of the i.f. impedance, the signal being amplified on a high-gain i.f. strip with subsequent detection. Calibration is effected using resistors in dummy crystal cartridges.

Admittance and v.s.w.r. measurements are carried out as for detectors, but usually it is the value presented to the local-oscillator source which is measured. For some purposes, it is desirable to use a more elaborate procedure in which the signal admittance is measured in the presence of local-oscillator excitation, and here one must employ the superheterodyne principle, the standing-wave carriage incorporating a head amplifier operating into an i.f. detection system.

With mixers used in pulse radars employing a common T–R system the ability to withstand leakage spikes of energy transmitted by a gas switch has been an important requirement in the past, although the situation is becoming modified with the advent of solid-state protective devices. Nevertheless the ability of mixers to withstand pulse burnout is still of interest in many applications. Short of using a high-power microwave test equipment powered by a magnetron, the use of a coaxial-line discharge system for burnout testing has proved most effective. An arrangement in which the switching is accomplished by means of a mercury-wetted relay operated at 100 Hz and giving pulse lengths of the order of $4 ns$ has been found to simulate well the "spike" leakage from gas T–R switches, mixers being required to withstand leakage energies of about 0.3×10^{-7} J per pulse.

Various automatic equipments have been developed to simplify the production testing of mixers such as units for displaying I/V characteristics and the direct display of conversion loss in a method dependent on the amplitude modulation of local-oscillator drive [9]. Probably the most important of such equipments, however, is an automatic noise-figure display unit which is illustrated schematically in Fig. 1.16.

In one version of this equipment developed at A.E.I. a noise tube acting as signal source is modulated by a square wave at a frequency of 375 Hz, and this signal is applied to the mixer along with the local-oscillator drive in a manner similar to that described for the manual system. The resulting i.f. signal is amplified by pre- and post-amplifiers which are this time connected directly together, and the 375 Hz component of the detected output is extracted and rectified. When applied to an output meter, assuming the gain of the system to be fixed, this will give a reading which is inversely proportional to the conversion loss of the mixer. If now a.g.c. is applied to the i.f. system in such a manner that the average level of the modulated wave at the detector output remains constant, this implies that the amplitude of the square wave is inversely proportional to the Y-factor of the mixer. The output meter reading derived from the amplitude of this modulation will then depend on the product of conversion loss and Y-factor, which, as indicated by eqn. (1.20), gives the overall noise figure.

Microwave Semiconductor Devices

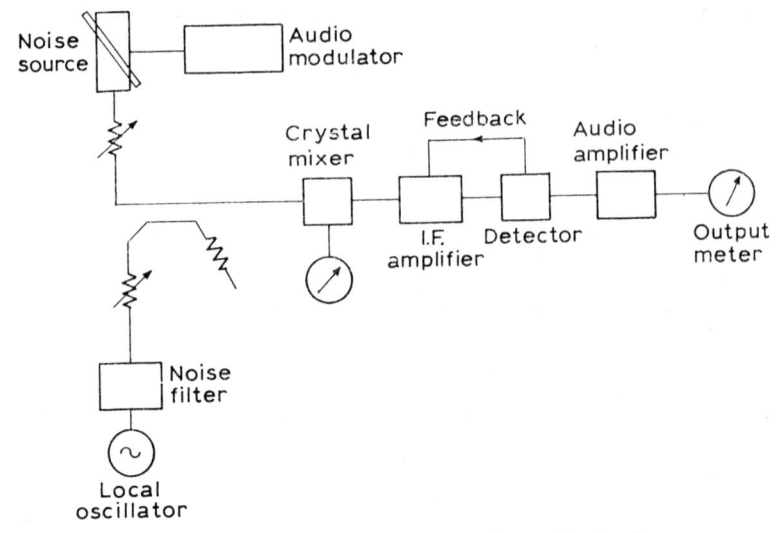

FIG 1.16 Equipment for automatic measurement of overall noise figure

The meter scale can therefore be calibrated to read overall noise factor directly.

Mixers for C.W. Radars

The most important measurement on mixers intended for Doppler applications is that of signal/noise ratio in a narrow frequency band of the order of 100 Hz at a frequency removed from that of the i.f. carrier by 1–10 kHz. A basic equipment is represented by the block diagram of Fig. 1.17, in which the amplitude-modulated noise on the i.f. carrier, converted into audio-frequency noise after detection, is examined on a spectrum analyser. For some applications frequency-modulated noise may be important also, but this we shall ignore.

The principal defect of the basic equipment is that noise from the klystron oscillators is of the same order as that arising in the crystals and limits the sensitivity of measurement. This method can only be used if specially selected klystron oscillators are available.

An alternative equipment, which largely overcomes the problem of noise in the klystrons, is based on a balanced mixer in which two crystals of opposite polarity are used in parallel. Both are fed with local oscillator and signal, so arranged that the i.f. outputs are in phase and may be

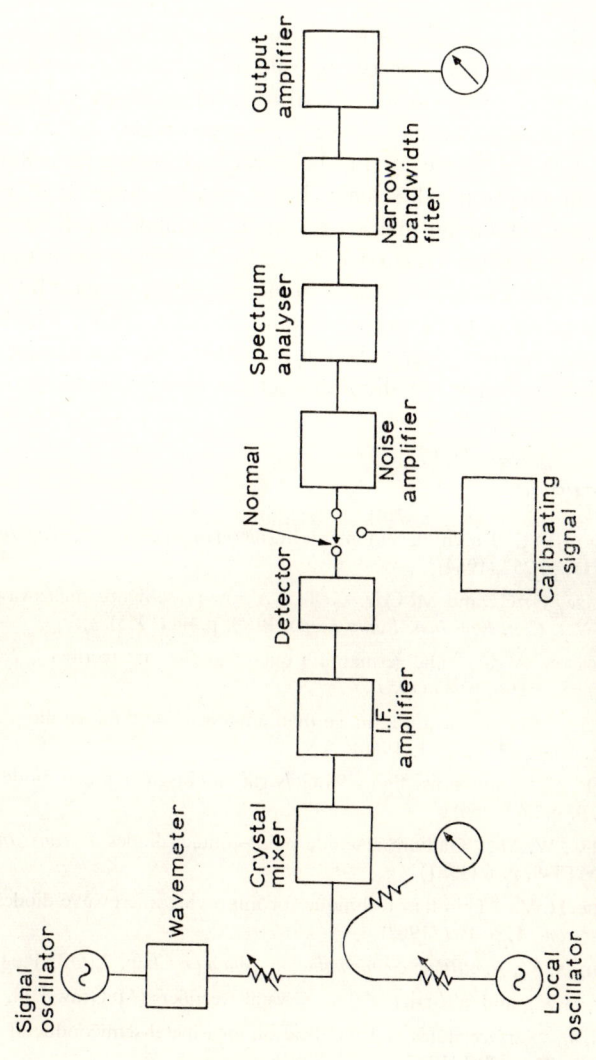

Fig 1.17 Flicker noise measuring equipment

introduced additively into the i.f. amplifier input impedance. The direct-current outputs, on the other hand, are of opposite polarity and therefore cancel, together with the i.f. component of noise which is attributable to the local-oscillator source.

Since the flicker-noise modulation with a balanced mixer is the sum of contributions from two crystals of opposite polarity, it is possible to utilize the feature that the phases of the signal or local oscillator voltages at the mixer may be changed so as to give i.f. outputs which are in antiphase and mutually cancel. Since mixing has occurred before cancellation the flicker noise modulation will remain, but it may be shown that, provided the signals are exactly in antiphase, the signal modulation will be cancelled to the same degree as the i.f. carrier. Thus, by cancelling the carrier and its associated modulation by say 20 dB and increasing the i.f. gain by the same amount, it is possible to obtain a measurable noise output which is pre-dominantly due to the mixer flicker modulation. The frontispiece shows such an equipment in use, the diodes under test being intended for the 2 cm band.

REFERENCES

1 SHURMER, H. V., "Recent developments in silicon radar crystals", *Proc. Instn Elect. Engrs*, **111**, p. 257 (1964).

2 MESSENGER, G. C., and MCCOY, C. T., "A low-noise-figure microwave crystal diode", *Nat. Con. Rec. Inst. Radio Engrs*, Pt. 8, p. 68 (1955).

3 MACPHERSON, A. C., "The germanium microwave crystal rectifier", *Trans. Inst. Radio Engrs*, **ED-6**, p. 83 (1959).

4 OXLEY, T., "Recent advances in germanium microwave mixer diodes", *Radio Electron. Engng*, **31**, p. 181 (1966).

5 BAUER, R. J., "A low noise figure 94 Gc/s gallium arsenide mixer diode", *Microwave J.*, **9**, p. 84 (1966).

6 SHARPLESS, W. M., "Gallium arsenide point-contact diodes", *Trans. Inst. Radio Engrs*, **MTT-9**, p. 6 (1961).

7 SHURMER, H. V., "Thin-film technique for improving microwave diodes", *Industrial Electron.*, **1**, p. 169 (1962).

8 HENISCH, H. K., *Rectifying Semiconductor Contacts*, Chap. 7 (O.U.P., 1957).

9 TORREY, H. C., and WHITMER, C. A., Crystal Rectifiers (McGraw-Hill, 1948).

10 BARDEN, J., "Surface states and rectification at a metal-semiconductor contact", *Phys. Rev.*, **71**, p. 717 (1947).

11 FROHMAIER, J. H., "Noise performance of a three-stage microwave receiver", *Electron. Tech.*, **37**, p. 245 (1960).

Point-contact Crystal Diodes

12 SHURMER, H. V., "Crystal detectors to cover the frequency band 26–40 Gc/s", *Proc. Instn Elect. Engrs*, **108**, p. 659 (1961).

13 SHURMER, H. V., "Mechanisms in silicon point-contact diodes", *J. Electron.*, **13**, p. 305 (1962).

14 SHURMER, H. V., "Noise performance of a three-stage microwave receiver", *Electron. Radio Engr*, **35**, p. 272 (1958).

15 JENNY, D. A., "A gallium arsenide diode", *Proc. Inst. Radio Engrs*, **46**, p. 717 (1958).

16 NICOLL, G. R., "Noise in silicon microwave diodes", *Proc. Instn Elect. Engrs*, **101**, p. 317 (1954).

17 SCHIFF, L. I., "Noise in crystal rectifiers", *NDRC* 14–126 (University of Pennsylvania, 1943) (summarized in Ref. 9, p. 187).

18 WEISSKOPF, V. F., "On the theory of noise in conductors, semiconductors and crystal rectifiers" (1943) (summarized in Ref. 9, pp. 179 *et seq.*).

19 SHURMER, H. V., "A crystal mixer for c.w. radars", *Proc. Instn Elect. Engrs*, **110** p. 117 (1963).

20 ENG, S. T., "A new low i.f. noise mixer diode: experiments, theory and performance", *Solid State Electron.*, **8**, p. 59 (1965).

21 ENGELSON, M., "Performance of harmonic diode mixers", *Microwaves*, **6**, p. 72 (1967).

22 MEREDITH, R., and WARNER, F. L., "Superheterodyne Radiometers for 70 Gc/s and 140 Gc/s". *Trans. IEEE*, **MTT-11**, p. 397 (1963).

23 BLEANEY, B., RYDE, J. W., and KINMAN, T. H., "Crystal valves", *J. Instn Elect. Engrs*, **93A**, p. 847 (1946).

24 ROBERTS, S., "Theory of noise measurements on crystals as frequency converters", R.L. Report No. 61-11 (1943) (summarized in Ref. 9, pp. 223 *et seq.*).

2

Varactor Diodes

2.1 INTRODUCTION

The non-linear barrier capacitance of a semiconductor diode, regarded as a parasitic element in mixers and detectors, is in fact the feature which in varactor diodes is utilized to form active or passive u.h.f. and microwave devices. We shall confine our attention in this chapter to varactor diodes which are intended for microwave use and which also employ p–n junction structures, although Schottky junctions, discussed in Chapter 3, can be used as varactors too. We should note that experimental varactors have also been made from insulating materials, principally silicon dioxide, using structures containing no rectifying barrier at all [1].

Widespread attention was first drawn in 1958 to the potential use of a new family of microwave devices based on the variable capacitance principle in a now classic paper by A. Uhlir [2], who attributed the coining of the name *varactor* to his colleague M. E. Hines. Uhlir indicated possible uses for these devices in low-noise amplifiers, amplifying frequency converters, harmonic and sub-harmonic generators, switches, limiters and voltage-tuned passive circuits, all of which have since become commonplace items.

The materials principally used for varactors are silicon and gallium arsenide, epitaxy* being widely used to reduce series resistance. The usual forms of construction as indicated in Fig. 2.1, although some devices offered as varactors employ point-contact structures which may operate as p–n junctions or as metal/semiconductor Schottky-type barriers.

The basic equivalent circuit for a varactor may be represented as in Fig. 2.2(*a*), on the assumption that the device is not driven into the forward conduction régime during its cycle of operation, in which case the effective

---

* *Epitaxy* is a process in which a semiconductor is grown from the vapour phase in the form of a layer of film.

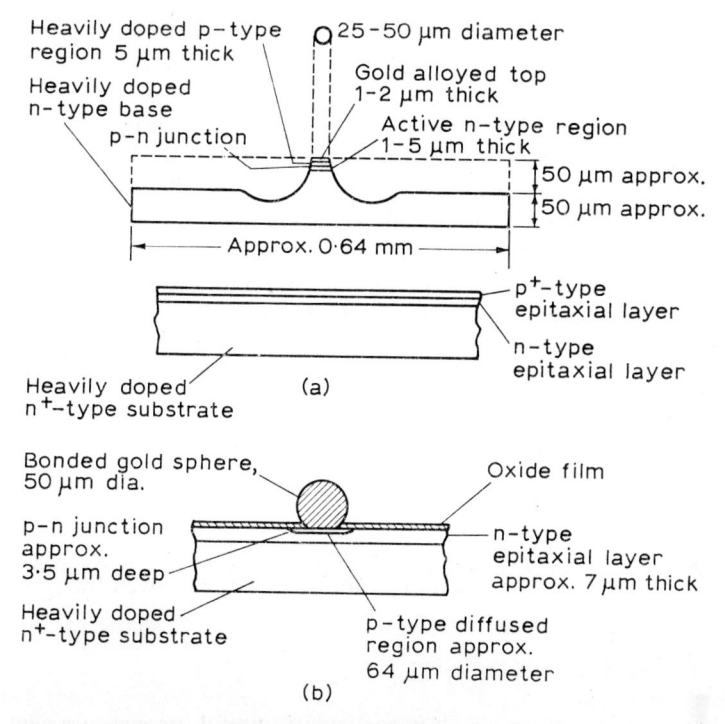

FIG 2.1. Typical epitaxial varactor structures
    (a) Mesa   (b) Planar

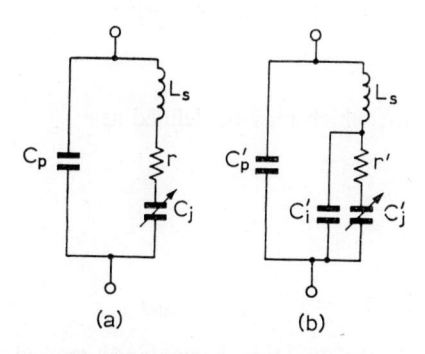

FIG 2.2 Equivalent circuits of varactor
    (a) Basic   (b) Modified

45

barrier resistance remains sufficiently high to exert only a negligible shunting effect on the barrier capacitance C_j. The value of series resistance r is so chosen as to account for all the loss associated with the junction; C_p represents the cartridge parallel capacitance plus strays. All the stray inductive reactance is assumed to be accounted for by the series inductance L_s.

The so-called "cut-off frequency" f_c gives an indication of loss in all applications but is not in itself a figure of merit. There can in fact be no universally applicable figure of merit for varactors since the uses are so diversified, but such a term may profitably be applied in relation to parametric amplifiers to the product of cut-off frequency and a non-linearity factor, as discussed below.

We define f_c as that frequency at which the capacitive reactance of the junction becomes equal to the series resistance, i.e.

$$f_c = \frac{1}{2\pi C_j r}$$

Since the Q-factor at any measuring frequency f is given by

$$Q = \frac{1}{2\pi f C_j r}$$

we have a relationship which is particularly useful for measurement purposes, namely

$$f_c = fQ$$

For parametric amplifiers, we define the figure of merit as

$$M = \gamma f_c = \gamma f Q$$

where γ is the non-linearity factor, which may be defined as

$$\gamma = \frac{\Delta C}{C}$$

Here

$$C = 2(C_{j\max} + C_{j\min}) \quad \text{and} \quad \Delta C = (C_{j\max} - C_{j\min})$$

where $C_{j\max}$ and $C_{j\min}$ represent the junction capacitance measured at $+1\,\mu\text{A}$ and $-1\,\text{V}$, respectively. A high figure of merit indicates a low

Varactor Diodes

attainable noise figure. At the optimum idling frequency the minimum predicted noise figure is given by [3]

$$F_{min} = 1 + \frac{2f_c}{M}$$

In practice it is difficult to separate the junction capacitance from that of the cartridge and strays, so that it is of interest to note the effect of C_p on the Q-factor and figure of merit. The new effective Q-factor is obtained by transforming the parallel circuit of Fig. 2.2(*a*) into an equivalent series circuit, when it is easily shown that the new value of Q is related to the intrinsic value for the junction by

$$Q' = Q\frac{C_j + C_p}{C_j}$$

and is therefore increased by stray capacitance. The non-linearity factor, on the other hand, is now given by

$$\gamma' = \frac{\Delta C_j}{C_j + \Delta C_j} = \frac{\Delta C_j}{C_j}\frac{C_j}{C_j + C_p} = \gamma\left(\frac{C_j}{C_j + C_p}\right)$$

Thus

$$\gamma'Q' = \gamma Q$$

which indicates that the figure of merit is unchanged by the presence of stray capacitance.

For parametric amplifiers, a breakdown voltage of a few volts will suffice and this allows the junctions themselves to be relatively heavily doped, thereby tending to reduce the series resistance. With harmonic multipliers, however, one is concerned with storing as much energy per cycle as possible, which suggests that the quantity $C_{jmin}V_B{}^2$ should be maximized, V_B being the breakdown voltage and C_{jmin} the associated minimum value of junction capacitance. Since C_{jmin} may be controlled independently of V_B, it is clearly desirable to make V_B high. With multipliers, high cut-off frequency, together with adequate charge storage and non-linearity, produces efficient harmonic generation.

In switching applications and for harmonic generators operating "step-recovery" or on the "snap-off" principle, fast transition times are required. It is difficult to generalize about the requirements here, as carrier lifetime, geometry, doping level and drive conditions are all involved. The ability of varactors to store charge through *diffusion capacitance* leads to multiplier diodes of much greater power capability than would be expected from the normal junction or *depletion capacitance*. As diffusion capacitance depends on a relatively long lifetime of carriers in the junction region, a feature known to be possessed by silicon, this was for some considerable time assumed to account for an apparent superiority of silicon over gallium arsenide in varactor frequency multipliers. Later it was reported that gallium-arsenide varactors could also be used successfully in these applications and the role of charge storage by diffusion capacitance then became somewhat obscure, a position which still appears unclarified.

It has been indicated that a high figure of merit for a varactor implies a low value for the Cr product, which was shown to be a requirement for good point-contact devices. This suggests that a good varactor material will have a high value for the product $\mu \epsilon^{-1/2}$, indicating why gallium arsenide is preferred for applications such as parametric amplifiers where requirements on breakdown voltage and carrier lifetime are not stringent.

2.2 THEORY OF OPERATION

2.2.1 The Abrupt *p–n* Junction

Let us consider an idealized situation in which the doping of the semiconductor changes suddenly from a uniform net acceptor concentration to a uniform net donor concentration, the impurity densities on the two sides of the boundary having unrelated arbitrarily assumed values, as indicated in Fig. 2.3(*a*). Junctions formed by alloying can approximate to such structures, and a metal/semiconductor junction has already been considered as one limiting case.

The energy band diagram and potential distribution will be as shown in in Fig. 2.3(*b*). Since the Fermi level in thermodynamic equilibrium must remain constant throughout the material, a potential energy difference eV_D is set up between *p*- and *n*-regions far removed from the junction, V_D being known as the *diffusion potential*.

We note that eV_D represents the difference in energy levels at either the bottom of the conduction bands W_C or the top of the valence bands W_V on the two sides of the junction. Relative to the Fermi levels W_F, each of these is determined by the thermal equilibrium density of holes p_0 or of electrons n_0 for the *p*- and *n*-sides of the junction, respectively, as follows:

$$\left. \begin{array}{l} n_0 = n_c\, e^{-(W_C - W_F)/kT} \\ p_0 = n_v\, e^{-(W_F - W_V)/kT} \end{array} \right\} \quad (2.1)$$

where

$$n_c = 2\left(2\pi m_e \frac{kT}{h^2}\right)^{3/2}$$

and

$$n_v = 2\left(2\pi m_h \frac{kT}{h^2}\right)^{3/2}$$

are known as the effective densities of states in the conduction and valence bands, respectively, representing the densities of states in a strip kT wide at the edge of each band. m_e and m_h are the effective masses of electrons and holes, respectively, and h is Planck's constant.

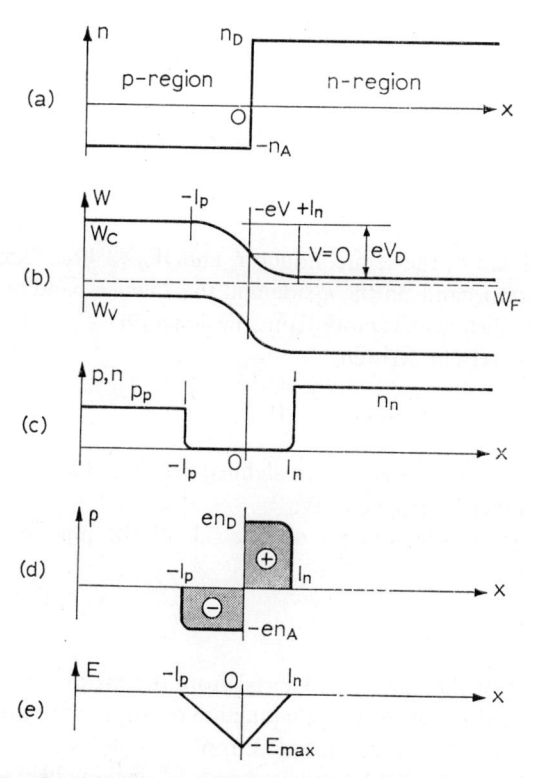

FIG 2.3 Abrupt p–n junction: detailed presentation
(From Ref. 4)

For silicon and germanium at room temperature,

$$n_c = 2{\cdot}5 \times 10^{25} \left(\frac{m_e}{m}\right)^{3/2} \text{ per cubic metre}$$

$$n_v = 2{\cdot}5 \times 10^{25} \left(\frac{m_h}{m}\right)^{3/2} \text{ per cubic metre}$$

where the ratio of the effective mass of an electron to its rest mass,

$$\frac{m_e}{m} = 0{\cdot}55 \text{ for germanium and } 1{\cdot}1 \text{ for silicon}$$

and the ratio of the effective mass of a hole to the rest mass of an electron,

$$\frac{m_h}{m} = 0{\cdot}37 \text{ for germanium and } 0{\cdot}57 \text{ for silicon}$$

The diffusion potential may be obtained from the equation

$$eV_D = W_G - (W_{C(n)} - W_{V(p)}) \qquad (2.2)$$

where W_G is the band gap of the semiconductor, and $W_{C(n)}$, $W_{V(p)}$ are the heights of the conduction band on the n-side and the valence band on the p-side, respectively, at distances remote from the junction. We note that eqn. (2.2) may alternatively be written

$$eV_D = W_G - (W_F - W_V)_p - (W_C - W_F)_n \qquad (2.3)$$

where $(W_F - W_V)_p$ and $(W_C - W_F)_n$ are defined by eqn. (2.1) for the p- and n-sides of the junction respectively.

If a metal replaces the semiconductor on one side of the junction, say the p-side, we then have

$$eV_D = e\phi_0 - (W_C - W_F)_n \qquad (2.4)$$

where $e\phi_0$ is the height of the barrier as viewed from the metal.

The change of potential V_D across the p–n junction results in the existence of a space-charge double layer, because electrons and holes are swept away from the boundary by the field associated with V_D, leaving behind the uncompensated acceptors and donors, as indicated in Figs. 2.3(c) and (d).

Varactor Diodes

Since the net space charge must be zero, the two shaded areas at (d) are equal. The electric field, shown at (e), is derived from Poisson's law

$$\text{div grad } \phi = -\text{div } E = -\frac{\rho}{\epsilon} \tag{2.5}$$

from which it follows that the field is a linear function of distance, its maximum value being given by

$$E_{\max} = -\frac{el_n n_D}{\epsilon} = -\frac{el_p n_A}{\epsilon} \tag{2.6}$$

We base the following analysis for an abrupt junction on the treatment given by Professor Jonscher in his *Principles of Semiconductor Device Operation* [4].

It is assumed that the space-charge regions of constant densities $-en_A$ and en_D extend to depths l_p and l_n into the *p*- and *n*-regions, respectively, assuming complete ionization of all acceptor and donor centres.

Thinking in terms of unidirectional current and arbitrarily choosing the voltage origin $V = 0$ at the boundary plane where the impurity concentration changes abruptly, we may write eqn (2.5) in the form

$$\frac{d^2\phi}{dx^2} = \frac{en_A}{\epsilon} \quad \text{for } -l_p < x < 0 \tag{i}$$

and

$$\frac{d^2\phi}{dx^2} = -\frac{en_D}{\epsilon} \quad \text{for } 0 < x < l_n \tag{ii}$$

If $-e\phi$ is the potential energy of an electron,

$$E = -\frac{d\phi}{dx}$$

The boundary conditions are

$$E = -\frac{d\phi}{dx} = 0 \quad \text{at } x = -l_p \text{ and } +l_n \tag{iii}$$

$$\frac{d\phi}{dx} \text{ is continuous at } x = 0 \tag{iv}$$

$$\phi(l_n) - \phi(-l_p) = V_D - V \tag{v}$$

Microwave Semiconductor Devices

where V is the applied forward voltage. The appropriate solutions of (i) and (ii) are

$$\phi = \frac{en_A}{2\epsilon}(x^2 + 2l_p x) \quad \text{for } -l_p < x < 0 \tag{vi}$$

$$\phi = \frac{en_D}{2\epsilon}(x^2 - 2l_n x) \quad \text{for } 0 < x < l_n \tag{vii}$$

It follows from condition (iv) that

$$n_A l_p = n_D l_n$$

and condition (v) gives

$$V_D - V = \frac{e}{2\epsilon}(n_D l_n^2 + n_A l_p^2) \tag{2.7}$$

Hence

$$l_p = (V_D - V)^{1/2}\left(\frac{2\epsilon}{e}\right)^{1/2}\left[\frac{n_D}{n_A(n_A + n_D)}\right]^{1/2} \tag{2.8}$$

$$l_n = (V_D - V)^{1/2}\left(\frac{2\epsilon}{e}\right)^{1/2}\left[\frac{n_A}{n_D(n_A + n_D)}\right]^{1/2} \tag{2.9}$$

The total depletion width D is thus given by

$$D = l_p + l_n = (V_D - V)^{1/2}\left(\frac{2\epsilon}{e}\right)^{1/2}\left[\frac{n_A + n_D}{n_A n_D}\right] \tag{2.10}$$

and the total depletion capacitance C over an area A is

$$C = \frac{A\epsilon}{D} = A\frac{\left(\frac{e\epsilon}{2}\right)^{1/2}\left(\frac{n_A n_D}{n_A + n_D}\right)^{1/2}}{(V_D - V)^{1/2}} \tag{2.11}$$

$$= C_0 \left(\frac{1}{1 - \frac{V}{V_D}}\right)^{1/2} \tag{2.12}$$

The electric field is found by differentiating eqns. (vi) and (vii):

$$E = -\frac{en_A}{\epsilon}(x + l_p) \quad \text{for } -l_p < x < 0 \tag{2.13}$$

$$= -\frac{en_D}{\epsilon}(l_n - x) \quad \text{for } 0 < x < l_n \tag{2.14}$$

Varactor Diodes

The maximum field may be written as

$$E_{max} = -2(V_D - V)^{1/2}\left(\frac{e}{2\epsilon}\right)^{1/2}\left(\frac{n_D n_A}{n_A + n_D}\right)^{1/2} = -\frac{2(V_D - V)}{l_p + l_n} \quad (2.15)$$

It is to be noted that when one side of the junction, say the *p*-side, has a much higher impurity concentration than the other, we arrive at the result already quoted for a metal/semiconductor contact, with eqn. (2.9) simplifying to

$$l_n \approx D \approx \left(\frac{2\epsilon(V_D - V)}{en}\right)^{1/2} \gg l_p \quad (2.16)$$

and eqn. (2.15) becoming

$$E_{max} = -2\left(\frac{(V_D - V)en_D}{2\epsilon}\right)^{1/2} = -\frac{2(V_D - V)}{l_n} \quad (2.17)$$

2.2.2 The Linear Graded Junction

A type of junction readily produced by the solid-state diffusion of impurities approximates to a linear impurity gradation. The reason may be understood by considering the formula appropriate to the diffusion of impurities across a boundary plane separating a region of semiconductor initially doped to a uniform concentration n_0 and a second region, initially undoped, i.e.

$$n = \frac{n_0}{2}\,\text{erfc}\,\frac{x}{2\sqrt{(Dt)}} \quad (2.18)$$

where x is measured from the boundary into the initially undoped region, D is the diffusion coefficient appropriate to the particular diffusion temperature employed, and t is the duration time for the diffusion.

It may be shown that, for distances from the boundary at which n has fallen by about two orders of magnitude, in practical cases being only a few tenths of a micron, the rate of decrease of n indicated by eqn. (2.18) becomes nearly exponential. We take the reference plane for a *p–n* junction as that at which the net acceptor or donor concentration is zero, i.e. $n_A = n_D = n_0{}^*$. A linear graded junction will then arise from the tail of a diffusion profile of ionized acceptor impurities intersecting one of ionized

53

Microwave Semiconductor Devices

donor impurities facing the opposite direction, as indicated in Fig. 2.4, in which case, close to the junction, we have approximately

$$n_A = n_0^* e^{k_1 x} \quad \text{and} \quad n_D = n_0^* e^{-k_2 x} \qquad (2.19)$$

where k_1 and k_2 are constants appropriate to the individual diffusion profiles, and x is the distance from the reference plane. The net acceptor concentration is then given by

$$n_A - n_D = n_0^*(e^{k_1 x} - e^{-k_2 x}) \approx n_0^*(k_1 + k_2)x \qquad (2.20)$$

provided that $(k_1 + k_2)x \ll 1$.

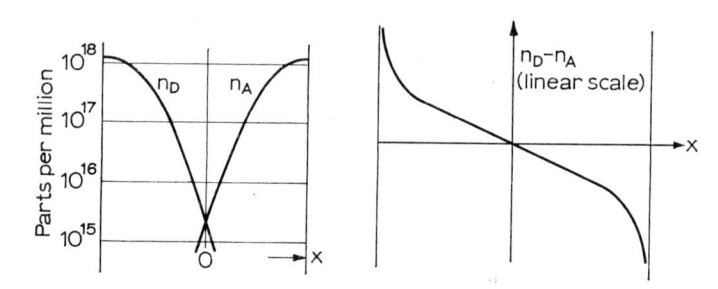

FIG 2.4 Linear graded junction: impurity distribution

Thus in the vicinity of the reference plane the net impurity concentration will change linearly with distance on either side.

Applying the same general principles to the linear junction as to the abrupt one, we may write Poisson's law as

$$\frac{d^2\phi}{dx^2} = \frac{-ea}{\epsilon} x \qquad (2.21)$$

where $a = n_0^*(k_1 + k_2)$. We find that the electric field is given by

$$E = \frac{ea}{2\epsilon}(x^2 - l^2) \qquad (2.22)$$

where the space-charge region now extends symmetrically about the reference plane from $-l$ to $+l$.

The potential ϕ is found by a second integration of eqn. (2.21):

$$\phi = \frac{ea}{6\epsilon}(3l^2 x - x^3) \qquad (2.23)$$

and the depletion width is given by

$$d = 2l = \left[\frac{12\epsilon(V_D - V)}{ea}\right]^{1/3} \qquad (2.24)$$

The depletion capacitance is therefore given by

$$C = A\left[\frac{\epsilon^2 ea}{12(V_D - V)}\right]^{1/3} \qquad (2.25)$$

$$= C_0 \left(\frac{1}{1 - \dfrac{V}{V_D}}\right)^{1/3} \qquad (2.26)$$

The maximum field, which occurs at the origin, is given by

$$E_{\max} = -\frac{ea}{2\epsilon}l^2 = -\left(\frac{9ea}{32\epsilon}\right)^{1/3} V_D^{2/3} = -\frac{3}{2}\frac{V_D}{2l} \qquad (2.27)$$

2.2.3 Other Forms of Junctions
Case 1

With alloyed/diffused *p–n* junctions (and also with some Schottky barriers) the impurity profile approximates to a heavily doped region on the one side of the junction reference plane (say the *p*-side) changing abruptly to a region of exponential (*n*-type) doping described by

$$n = n_0^* \, e^{kx}$$

The depth l_p to which the space-charge region extends into the heavily doped region will then be so small that it may be neglected in comparison with l_n, i.e. $l_n \approx D \gg l_p$. Poisson's law is now expressed as

$$\frac{d^2\phi}{dx^2} = \frac{-en_0^*}{\epsilon} e^{kx} \qquad (2.28)$$

and the electric field is given by

$$E = \frac{en_0^*}{\epsilon k}(e^{kx} - e^{kd}) \qquad (2.29)$$

The boundary conditions are taken to be

$$\phi = V_D - V \quad \frac{d\phi}{dx} = 0 \text{ at } x = d$$
$$\phi = 0 \text{ at } x = 0$$

Hence

$$V_D - V = \frac{en_0^*}{\epsilon k}\left[e^{kd}\left(d - \frac{1}{k}\right) + \frac{1}{k}\right] \tag{2.30}$$

which may be solved numerically for d and hence C evaluated.

Case 2

It sometimes happens that, for the types of junction considered in Case 1, the impurity distribution on the side which was taken as *n*-type approximates more nearly to a quadratic than to an exponential, i.e.

$$n = a + bx + cx^2$$

We then have

$$\frac{d^2\phi}{dx^2} = -\frac{e}{\epsilon}(a + bx + cx^2) \tag{2.31}$$

Taking the same boundary conditions as previously the electric field is found to be

$$E = \frac{e}{\epsilon}a(x - d) + \frac{b}{2}(x^2 - d^2) + \frac{c}{3}(x^3 - d^3) \tag{2.32}$$

A second integration leads to

$$V_D - V = \frac{e}{\epsilon}\left(\frac{ad^2}{2} + \frac{bd^3}{3} + \frac{cd^4}{4}\right) \tag{2.33}$$

from which d and therefore C may be found.

2.2.4 Capacitance Response to Fast Voltage Change

With varactors used as active devices at microwave frequencies, it is necessary that the depletion capacitance shall respond very rapidly to changes in applied voltage [5]. This depends on the dielectric relaxation time τ_d, which is usually defined in relation to the dispersion of a local excess of charge introduced into a dielectric medium. This τ_d is the time-constant ϵ/σ in the decay formula

$$q = q_0\, e^{-(\sigma/\epsilon)t}$$

Consider a varactor having an asymmetrical abrupt *p–n* junction, heavily doped on the one side and more lightly on the other, so that the depletion layer thickness is given by eqn. (2.16), which for present purposes we will write as

$$d = \left(\frac{2\epsilon}{en}\right)^{1/2} (V_s - V)^{1/2}$$

where V_s represents the static voltage (diffusion voltage plus bias voltage), and V is the instantaneously variable component. Suppose that an incremental voltage δV is suddenly applied across the varactor, the thickness of whose semiconducting element will be taken as w. For a one-dimensional representation, the incremental field δE due to δV initially set up across this element will be given by

$$\delta E = \frac{\delta V}{w}$$

The velocity with which the carriers at the edge of the depletion layer begin to respond to this field is

$$u_0 = \frac{\mu \, \delta V}{w}$$

Let δd be the change in depletion layer thickness corresponding to δV when equilibrium has been reached. The velocity will fall nearly exponentially to zero as this condition is approached, and the time-constant is obtained by assuming that the initial velocity is maintained, i.e.

$$t = \frac{w \cdot \delta d}{\mu \, \delta V} \approx \frac{w}{\mu} \frac{d(d)}{dV} \tag{2.34}$$

Hence, from eqn. (2.16),

$$t = \frac{w}{d} \frac{\epsilon}{\mu e n} = \frac{w}{d} \tau_d \tag{2.35}$$

We should note that the same result can also be derived from the equivalent electrical circuit.

For silicon varactors $\epsilon = 12 \times 8\cdot 85 \times 10^{-12}$ F/m and $\sigma \approx 1\,000$ S/m, so that $t \approx 10^{-13}$ s for $d \approx w$. Thus the varactor depletion layer can respond to a suddenly applied increment of voltage in a time which is short compared to the period of a microwave.

In varactors used for parametric amplifiers and with harmonic generators, the driving voltage is relatively large and it is of importance to consider how rapidly the depletion layer can respond to such excitation. If the drift velocity corresponding to the time derivative of the depletion layer formula becomes of the same order as the thermal velocity of the charge carriers, the assumption of virtually instantaneous response will not necessarily be justified.

For a varactor driven from a high-level sinusoidal voltage source the general solution is complicated when allowance is made for the resistance drop across the neutral bulk of the semiconductor behind the depletion layer. However, assuming that the amplitude of the driving voltage is made equal to the total static voltage V_s, a rough solution may readily be obtained for the maximum velocity of the depletion layer boundary, by neglecting the resistance drop, the result being

$$\left[\frac{d(d)}{dt}\right]_{max} = \omega \left(\frac{V_s}{2en}\right)^{1/2} \qquad (2.36)$$

For a typical high-frequency varactor driven at 10 GHz, assuming that $V_s \approx 5\,\text{V}$, and $d_0 \approx 10^{-7}$ m, the above expression gives

$$\left[\frac{d(d)}{dt}\right]_{max} \approx 10^4\,\text{m/s}$$

which is less than the thermal velocity of carriers by an order of magnitude.

2.2.5 Storage Capacitance

At microwave frequencies a substantial amount of charge may be represented by intermingled holes and electrons, and this leads to carrier storage capacitance in addition to the depletion capacitance already discussed. Following Uhlir [2], we illustrate how this arises for the particular case of a linear-graded junction.

For a typical graded junction at zero bias the concentrations p of holes and n of electrons vary with distance as shown in Fig. 2.5(a). The net charge density is $e(n_{D+} - n_{D-})$, taking n_{D+} and n_{D-} as the concentrations of ionized acceptors and donors respectively. The depletion, or space-charge, layer is centred, as we have previously seen, about the place where the net charge density is zero; this is known as the *stoichiometric junction*. There are very few holes or electrons within the depletion layer, but outside it the mobile carriers are present in almost exactly the right numbers to neutralize the fixed charges, the concentrations of holes and electrons varying with distance as shown in Fig. 2.5(a).

Varactor Diodes

Application to the contact of a slight forward bias will urge the hole and electron distributions towards each other, thereby decreasing the thickness of the depletion layer in accordance with the formula previously derived. The situation is now as illustrated in Fig. 2.5(*b*). Additional charges must at the same time enter at the contacts in order to maintain charge neutrality outside the depletion layer, and it is assumed that the contacts are of such a nature as to permit this.

On applying a still larger forward voltage, holes and electrons can intermingle for a time in a not-so-thoroughly depleted layer and on either

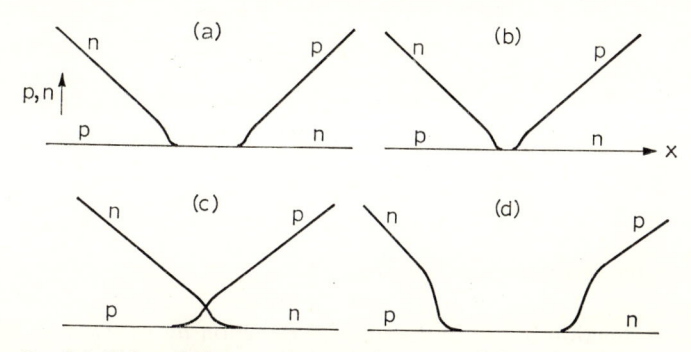

Fig 2.5 Hole and electron concentrations at graded *p–n* junction
(*a*) Zero bias
(*b*) Small forward bias
(*c*) Large forward bias
(*d*) Reverse bias
(*From Ref.* 2)

side of it, as illustrated in Fig. 2.5(*c*). Despite the intermingling, the charge may still be recovered if it is allowed to return in a time short compared with that for appreciable recombination. This situation does in fact usually apply if the frequency is much higher than 1 MHz.

For reverse bias, which widens the depletion layer, the situation is of the form illustrated in Fig. 2.5(*d*) until the bias reaches the breakdown voltage, at which carriers are generated copiously by avalanche multi-plication. In the range of voltage between breakdown and slight forward bias represented by Fig. 2.5(*b*) the effective capacitance is that due to the space-charge effect alone, i.e. the depletion capacitance.

2.2.6 "Step-recovery" Diodes

The removal of charge stored by forward current in a *p–n* junction device can lead to a very fast interruption of current, and this may be utilized

Microwave Semiconductor Devices

to give ultra-fast switching action, waveshaping or to generate harmonics well into the microwave spectrum. Such diodes have been termed "step-recovery" or "snap-off" types, according to the manufacturer. The abrupt transition is obtained by so designing the diode that, when the majority carriers at the junction are depleted by reverse bias, essentially all the minority carriers also have been removed. These junctions are built with retarding fields for the minority carriers so that storage is confined to the immediate vicinity of the junction. Their decay time may be several orders of magnitude less than the carrier lifetime. The following treatment of the storage and transition phases is due to Moll [6].

Storage Phase

The current flowing into the diode supplies the recombination current and increases the stored charge in accordance with the equation

$$i = \frac{dq}{dt} + \frac{q}{\tau} \tag{2.37}$$

where τ is the recombination lifetime.

We will restrict ourselves to the consideration of rectangular waveforms and take a junction biased in the forward direction by a current I_f. If the duration of current is several times the lifetime τ, the steady-state value of charge q_0 will be reached; otherwise it will be necessary to solve eqn. (2.37) for the stored charge at the beginning of the reverse transient. If at $t = 0$ the current is reversed to $-I_r$, the storage time is

$$t_s = \tau \log_e \left(1 + \frac{q_0}{I_r}\right) \tag{2.38}$$

where $q_0 = I_f \tau$ if the forward current has been on for a time $t_f \gg \tau$. In general, the initial charge is

$$q_0 = I_f \tau (1 - e^{-t_f/\tau}) \tag{2.39}$$

where t_f is the period of forward conduction.

Equation (2.38) slightly overestimates the storage time, since there is a finite residual charge left in the junction at the end of the storage phase, as illustrated in Fig. 2.6.

It should be noted that free-charge lifetime is not a fixed constant for a diode and may increase by some 50% for a 70°C temperature rise. Junction size is also important, smaller diodes tending to have a lower lifetime

Varactor Diodes

as well as a smaller temperature coefficient, because of increased ratio of surface to bulk recombination. At high stored charge the lifetime may be reduced by some 50% because of non-linear recombination and also on account of some of the stored charge extending into regions where the

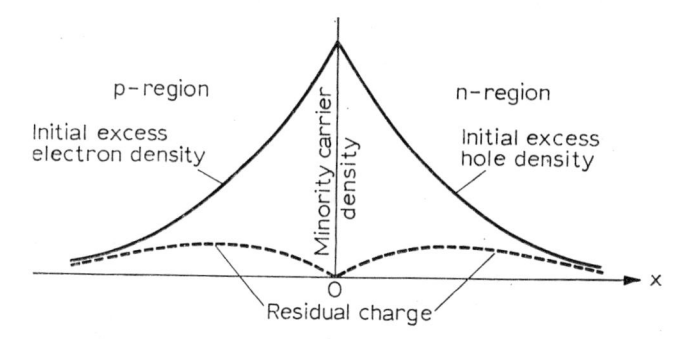

FIG 2.6 Initial distribution of excess charge and residual charge at end of storage phase
(*From Ref. 6*)

lifetime is shorter. At very low charge values the lifetime may also be reduced because of traps remaining unfilled, and at high impurity gradients there is a tendency for the lifetime to decrease.

Transition Phase
In diodes with retarding fields, the reverse current is due to both diffusion and drift, whereas with a step junction all the current is by diffusion. At the end of the storage phase, however, substantially all of the current must be due to diffusion, since both hole and electron densities of the junction approach zero. We can therefore calculate the gradient of carrier density and the junction, making the following analytical approximation to the residual carrier density, which Moll indicates can usually be justified:

$$\left. \begin{array}{l} p = \dfrac{I_{r(holes)}}{eD_pA} \, e^{-x/x_0} \quad x > 0 \\[1em] n = \dfrac{I_{r(electrons)}}{eD_nA} \, e^{+x/x_0} \quad x < 0 \end{array} \right\} \quad (2.40)$$

where $x_0 = kT/q(E)$; (E) is the average field; D_p, D_n are diffusion constants for holes and electrons, respectively; and A is the junction area.

We will make further assumption that for a symmetrical junction the electron and hole currents are equal, in which case

$$I_{r(holes)} = I_{r(electrons)} = \tfrac{1}{2} I_r$$

The fall of reverse current is a very complicated function of time, but since we are interested only in the approximate cut-off transition time, we assume for simplicity an exponential decrease:

$$i_r(t) = I_r \, e^{-t/t_q} \tag{2.41}$$

Taking the cut-off transition time to be small in comparison with the lifetime, we neglect the recombination loss of carriers during the cut-off phase, thereby obtaining

$$\int_0^\infty i(t)\,dt = \Delta q_p + \Delta q_n \tag{2.42}$$

where $\Delta q_p + \Delta q_n$ is the residual charge of electrons and holes, and time is measured from the end of the storage phase. For equal electron and hole currents, it follows from eqn. (2.40) that

$$I_r t_q = \frac{I_r x_0^2}{2}\left(\frac{1}{D_p} + \frac{1}{D_n}\right) \tag{2.43}$$

or

$$t_q = \frac{x_0^2}{D} \quad \text{where } D = \frac{2 D_p D_n}{D_p + D_n} \tag{2.44}$$

If $D_p = D_n$ we note that the decay time reduces to the diffusion time from the original centre of gravity of the injected carriers to the junction. Moll shows that, in this case,

$$t_q = \frac{1}{D}\left[\frac{q_0/A}{(4e)(b)(n_0 b)}\right]^{2/3} \tag{2.45}$$

where n_0, b are defined by the approximate relationships $n_D - n_A = 2n_0 \sinh bx$ and $b \approx w/4Dt$, with $w =$ wafer thickness, $D =$ diffusion constant and $t =$ diffusion time.

It is to be noted that $n_0 b$ is the impurity gradient in the junction and largely controls the breakdown voltage, sharper gradients being associated with shorter decay and lower breakdown. For a constant breakdown, the decay time decreases with increase in b, which in practice may be accomplished by using a very thin wafer and diffusing for a relatively short time.

Varactor Diodes

Reduction of the stored charge per unit area will also shorten the decay time, and a compromise must be made between the latter and the diode capacitance. The passive circuit parameters must also be optimized to give the maximum turn-off rate.

2.2.7 "Punch-through" Varactors

It has been claimed by one American manufacturer (Sylvania) that improved varactors for multiplier applications can be made by employing the "punch-through" technique, the comparison between this type and the more usual epitaxial varactors being indicated in Fig. 2.7.

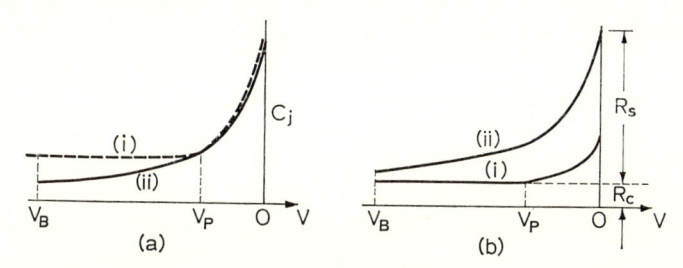

FIG 2.7 Comparison of "normal" and "punch-through" varactors
  (i) "Punch-through', varactor, $|V_P| < |V_B|$
  (ii) "Normal" varactor, $|V_B| < |V_P|$
  (*Sylvania Electric Products*)

Depending upon the thickness and resistivity of the junction layer, the space-charge layer may or may not have reached through to the far side of the junction layer at the breakdown voltage. With the "punch-through" varactor this condition is reached at a voltage which is low compared to the breakdown voltage and the diode is normally operated well beyond the punch-through voltage. The main advantage claimed for these varactors is that of higher conversion efficiency, ascribed to lower effective values of series resistance, discussed below.

Referring to a p^+–n–n^+ junction, the series resistance R_s consists of the sum of four terms: $R_s = R_p + R_n + R_{n+} + R_c$ where the first three terms relate to the p-, n- and n^+-layers respectively, and R_c represents the ohmic contact resistances.

In practice, R_c is usually a few tenths of an ohm at ultra-high frequencies but may be higher at microwave frequencies because of skin effect. For surface concentrations which are usual in epitaxial varactors, R_p will be negligible in comparison with R_c and R_n. Similarly R_{n+} will also be negligible for a highly doped substrate. The dominant component of R_s is therefore R_n, given by $R_n = \rho l/A$, where $l = w - d$, w being the width of the n-layer, d the depletion layer thickness and ρ the resistivity.

Since d is voltage dependent, so also are R_n and R_s. As with the junction capacitance C_j, if $|V_B| < |V_P|$ then R_s is a continuously decreasing function as voltages from zero to V_B are applied. If $|V_B| > |V_P|$ then in an ideal situation R_n will vanish at V_P, so that R_s will become very small and approximately equal to R_c, as indicated in Fig. 2.7(b). The change in series resistance with reverse voltage may in practice be appreciable, of the order of 2:1 for varactors with V_B in the range 50–100 V, and up to 10:1 for higher-voltage varactors. A "punch-through" varactor which has the same series resistance as a conventional varactor at breakdown will tend to have a lower value at small reverse voltages so that the average of R_s over the drive cycle will be lower.

2.2.8 Effect of Drive on Q-factor

Since the Q-factor gives the ratio of average energy stored in a capacitor to the energy dissipated in the effective series resistance per cycle its value must be of interest in all varactor applications, so that it is important to know how the Q-factor varies with the driving voltage. For applications which involve the small-signal depletion capacitance $C(V)$ and conductance $G(V)$ we note that these are represented in terms of the bias voltage by

$$C(V) = K(V_D - V)^{-n} \tag{2.46}$$

$$G(V) = \frac{e}{kT} I_0 \, e^{eV/kT} \tag{2.47}$$

where n is $\tfrac{1}{2}$ for abrupt junctions, and $\tfrac{1}{3}$ for linear graded junctions.

The general shapes of these functions are indicated in Figs. 2.8(a) and (b), and the resulting shape of the Q-factor curve for a conductance in parallel with a capacitance is shown at (c), where

$$Q_P = \frac{\omega C(V)}{G(V)}$$

defined on a small-signal basis [7].

The difference between the maximum forward voltage and the contact potential ϕ is denoted by V_R, and Q_P usually falls off rapidly near the voltage $\phi - V_R$, as indicated in Fig. 2.8(c). The limitation imposed by this

Varactor Diodes

decrease in Q_P is immediately apparent in relation to varactors used as circuit tuning elements, since these should in general not degrade the Q-factor of the fixed tuned circuits by more than, say, a factor of two. In other applications, where the varactor is a.c. driven, a slight excursion

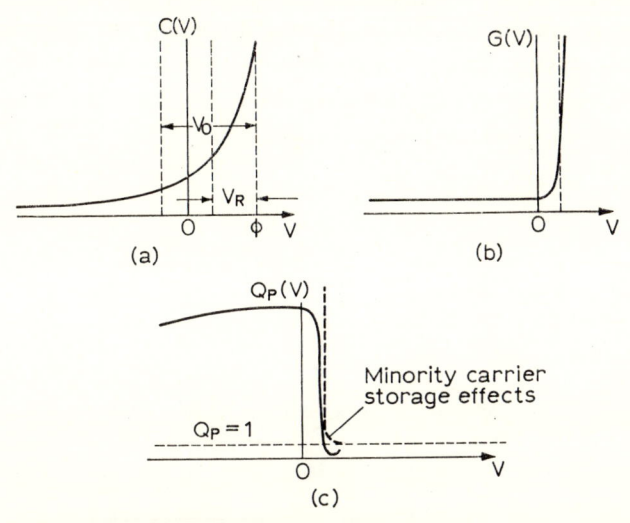

FIG 2.8 General varactor diode parameters
(*From Ref. 7*)

into the low-Q régime is permissible, provided that this occupies only a small fraction of a cycle, and a steady bias may be used to this end. The principal advantage is that the capacitance variation is thereby appreciably increased.

2.3 APPLICATIONS

In the following sections we will give a brief indication of the manner in which varactors are principally used. It will be appreciated that varactor applications forms a subject in its own right and we can here attempt no more than to highlight a few of the more general principles.

2.3.1 Parametric Amplifiers

In a parametric amplifier, the energy at one frequency is increased by supplying energy at a second frequency; the basic idea is illustrated in

Fig. 2.9. Considering the simple resonant circuit of Fig. 2.9(a), it will be assumed for the purpose of illustration that the plates of the capacitor can be separated mechanically. Suppose that prior to a time $t = 0$ the circuit has been induced to oscillate at its natural frequency and that upon the capacitor reaching its first maximum charge after $t = 0$ the separation of the plates is suddenly increased, thereby decreasing the capacitance. Because of the electric field, this separation requires mechanical work to

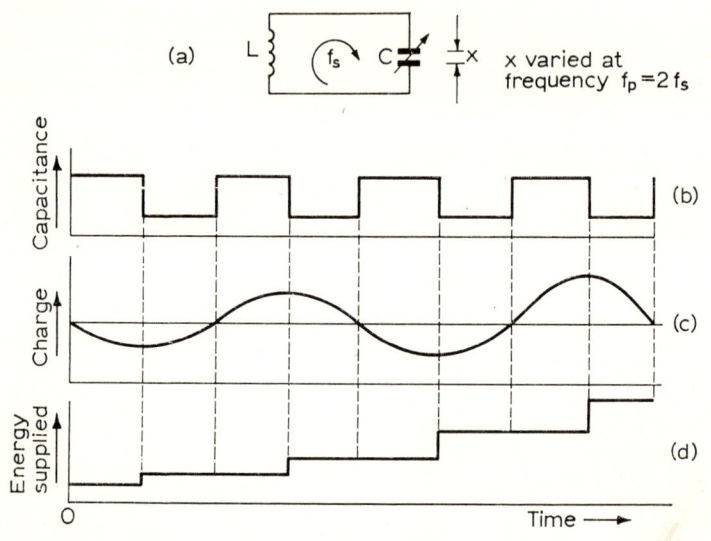

FIG 2.9 Illustration of basic parametric amplifier principles
(From Ref. 26)

$C = \dfrac{\varepsilon A}{d}$

be supplied, and this is accompanied by an equally sudden increase in voltage. The relation between charge q, capacitance c and voltage v on a capacitor is, of course, $q = cv$. Thus if the plate separation is increased and the capacitance thereby decreased, in a time short enough for q to be considered essentially constant, the voltage and also the stored energy $\frac{1}{2}qv$, must both be increased. If at the next zero of voltage across the plates the original separation is suddenly restored this does not change the stored energy of the circuit since the field is then zero, so that, if the whole sequence is then repeated giving the capacitance variation shown in Fig. 2.9(b), the stored energy may be increased every half-cycle, giving a build-up of charge across the capacitor as indicated at (c) and an increase in energy as shown at (d). This build-up will go on until the energy added per

Varactor Diodes

separation of the plates is just equal to the energy dissipated per cycle. It is to be noted that the plates are *pumped* at twice the resonant frequency and that the phase of the pumping is important. A variable-capacitance amplifier operating exactly on this principle, in which a signal is supplied at the resonant frequency f_s and the pump frequency $f_p = 2f_s$, is said to be *degenerate*.

Non-degenerate parametric amplifiers, which do not impose the above restriction on pump frequency or phase, can be constructed by connecting

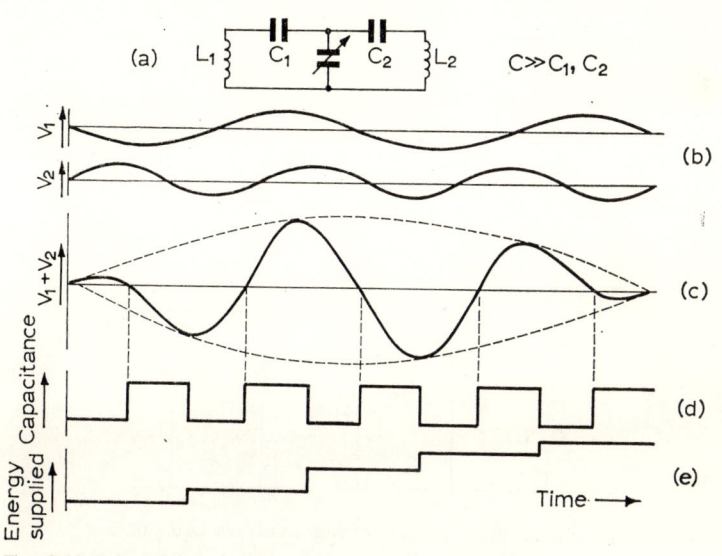

FIG 2.10 Principle of non-degenerate parametric amplifier
(*From Ref.* 26)

a second tuned circuit across the variable capacitance, as shown in Fig. 2.10(*a*). This circuit is known as the *idler* and its frequency f_i is tuned to $f_p - f_s$, the reason for which is indicated below.

Assuming unity amplitude in each case, let the charge associated with f_s be $\sin \omega_1 t$ and that associated with f_i be $\sin \omega_2 t$. The total charge is then

$$q = \sin \omega_1 t + \sin \omega_2 t$$

$$= \sin \left(\frac{\omega_1 + \omega_2}{2} + \frac{\omega_1 - \omega_2}{2} \right) t + \sin \left(\frac{\omega_1 + \omega_2}{2} - \frac{\omega_1 - \omega_2}{2} \right) t$$

$$= 2 \sin \left(\frac{\omega_1 + \omega_2}{2} \right) t \cos \left(\frac{\omega_1 - \omega_2}{2} \right) t \qquad (2.48)$$

Microwave Semiconductor Devices

We note that the charge becomes zero every time $\sin 2\pi \times \frac{1}{2}(f_1 + f_2)t$ has a zero, so that the possibility of building up oscillations exists just as in the degenerate case. The voltage which would appear across either tuned circuit if the other were removed is shown in Fig. 2.10(*b*), and the combined voltage is given at (*c*). The necessary capacitance variation due to pumping is shown at (*d*), and the resulting energy build-up at (*e*). In this case the energy increments vary considerably with the phase of the beat frequency but the net energy transfer is always positive.

Manley–Rowe Equations

In considering the action of parametric amplifiers, a powerful tool lies in a result originally derived by Manley and Rowe [8]. It is convenient for the purpose of illustration to consider the general situation as represented by the circuit of Fig. 2.11, in which each frequency flows in a

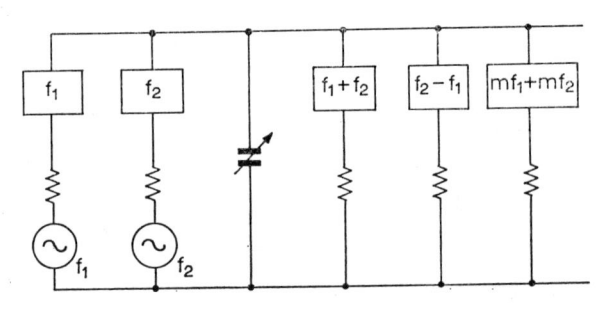

FIG 2.11 Circuit illustrating use of Manley–Rowe equations

separate external circuit. Each of these circuits contains in series a voltage generator of the appropriate frequency, a load impedance and an ideal filter, which presents a short-circuit to the desired frequency and an open-circuit to all other frequencies. The voltages of all generators in this equivalent circuit, except those corresponding to the sources driving the non-linear capacitor, are set equal to zero.

The simplest derivation of the Manley–Rowe equations is probably that based on a quantum argument, due to M. T. Weiss [9], which is as follows. Consider a non-linear reactance connecting energy sources and sinks at a set of frequencies $mf_1 + nf_2$, where m and n are integers and f_2/f_1 is irrational. If the power introduced into the reactance is P_{mn} at a frequency $mf_1 + nf_2$, it can be considered as produced by X_{mn} quanta

Varactor Diodes

per second, each quantum carrying energy $h(mf_1 + nf_2)$, h being Planck's constant. Thus

$$P_{mn} = hX_{mn}(mf_1 + nf_2) \tag{2.49}$$

Since no power is absorbed in the reactance it follows that

$$\sum_{m,n} P_{mn} = \sum_{m,n} hX_{mn}(mf_1 + nf_2) = 0 \tag{2.50}$$

where the values of m and n are taken for all sources and sinks. From this equation,

$$\sum_{m,n} mX_{mn} + \frac{f_2}{f_1}\sum_{m,n} nX_{mn} = 0 \tag{2.51}$$

Since X_{mn}, m and n are integers and f_2/f_1 is irrational, eqn. (2.51) can be satisfied only if the two parts are separately equal to zero, giving

$$\sum_{m,n} mX_{mn} = \sum_{m,n} nX_{mn} = 0 \tag{2.52}$$

Substitution for X_{mn} from eqn. (2.49) gives the Manley–Rowe equations, namely

$$\sum_{m,n} \frac{mP_{mn}}{mf_1 + nf_2} = 0 \tag{2.53}$$

$$\sum_{m,n} \frac{nP_{mn}}{mf_1 + nf_2} = 0 \tag{2.54}$$

Applications of the Manley–Rowe equations to some practical situations in which power is permitted to flow at three frequencies only is illustrated in the following examples, taking the non-linear reactance to be that of a capacitor.

Example 2.1

If power is allowed to flow at only the signal frequency f_1, the pump frequency f_2 and the sum frequency $f_3 = f_1 + f_2$, then eqn. (2.53) reduces to

$$\frac{P_1}{f_1} + \frac{P_3}{f_3} = 0 \tag{2.55}$$

and eqn. (2.54) gives

$$\frac{P_2}{f_2} + \frac{P_3}{f_3} = 0 \tag{2.56}$$

As the pump power P_2 supplied to the capacitor is positive, it follows from eqn. (2.55) that the power at the sum frequency f_3 is negative and therefore flows from the capacitor into the load. Secondly, defining the gain G_{13} as the ratio of power at f_3 flowing into the load to power at f_1 into the reactance gives

$$G_{13} = \frac{f_3}{f_1} \tag{2.57}$$

which represents the maximum gain of the sum-frequency amplifier or up-converter irrespective of circuit configuration or degree of non-linearity.

Example 2.2
Suppose now that power is allowed to flow at a frequency which is the difference between the pump and signal frequencies, i.e. the pump frequency is now represented by f_3 and the output frequency by f_2. Since power in this case is being supplied at f_3 it follows that both P_1 and P_2 are negative, and hence the varactor supplies power to the source at f_1 rather than absorbing it. Since this power is independent of that supplied by the source itself, it follows that infinite gain is possible and the device is capable of oscillation at both f_1 and f_2. This is the situation which obtains in the negative-resistance parametric amplifier.

Negative-resistance Amplifier
The negative-resistance amplifier is probably the most popular of the parametrics, and we here use this type as an example to illustrate their principal features. The basic circuit is shown in Fig. 2.12(*a*).

Parametric amplifiers are essentially low-noise devices because, at least in the ideal case, there is no resistance and no conduction current. This situation is in contrast to transistor amplifiers, for example, which inherently produce both thermal noise and shot noise. Assuming that non-linear capacitors are noiseless devices, then the Manley–Rowe equations and derived expressions can be applied to them, with the reservation that there will in practice always be some resistive component and it will be necessary to make a real circuit analysis to obtain exact expressions for gain and noise figure.

In low-noise amplifiers, it is common practice to express noise performance in terms of effective noise temperature, which is related to the noise figure F by the equation

$$T_e = T_0(F - 1) \tag{2.58}$$

where T_0 is regarded as room temperature, standardized at 290 K.

It may be shown [10] that for negative-resistance amplifiers the normalized effective temperature is given by

$$\frac{T_e}{T_0} = \frac{r}{R_g} + A\frac{f_1}{f_2}\frac{(r + R_g)}{R_g} \tag{2.59}$$

where r is the series resistance of the varactor, R_g the source resistance, A the ratio of negative resistance to the total positive resistance in the signal circuit, f_1 the signal frequency, and f_2 the idling frequency. R_g is

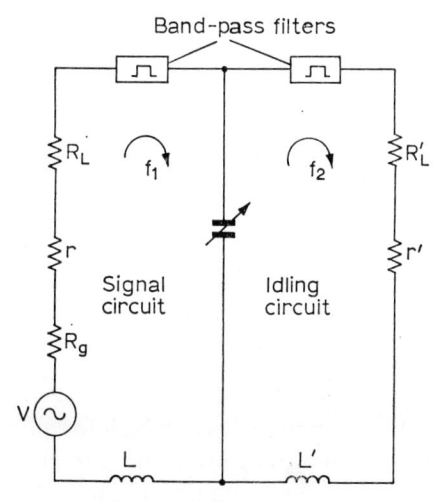

Fig 2.12 Basic circuit for negative-resistance amplifier

R_L = load resistance
r = internal resistance
R_g = source resistance

selected so that the diode current due to pumping is small, typically 1 μA, in order that shot noise associated with this current should remain negligible. With parametric amplifiers the incoming and outgoing signals may be separated by means of a non-reciprocal multi-port ferrite device known as a *circulator*. If a circulator is used, this will have a finite isolation and an insertion loss, so that eqn. (2.59) will have to be modified accordingly. Calculated effective noise temperatures are of the order of 80 K for a signal frequency of 3 GHz, rising to some 230 K at 8 GHz. Measured values not many degrees in excess of these figures have been reported [11].

Microwave Semiconductor Devices

The negative conductance may be obtained using an analysis given by Heffner and Wade [12], as follows. Consider a two-tank* circuit amplifier as shown in Fig. 2.13. We assume that the source is tuned to the resonant frequency ω_1 of tank 1 producing a voltage $V(\omega_1)$, and that mixing in the varactor produces a voltage $V(\omega_2)$ across the idler circuit, tank 2, tuned to ω_2. The polarity of voltages across both tanks and the positive directions

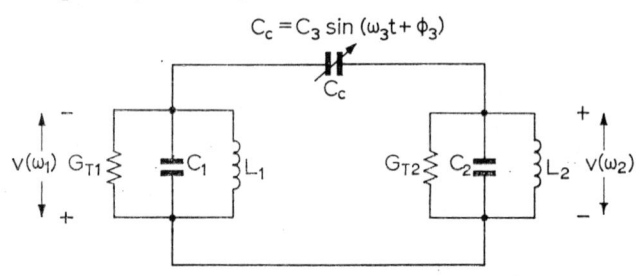

FIG 2.13 Equivalent circuit of two-tank amplifier
(*From Ref.* 12)

of current are indicated by arrows. G_{T1} and G_{T2} designate the internal shunt conductances of the tanks.

The voltage appearing across the varactor may be expressed as

$$v_c = V_1 \sin(\omega_1 t + \phi_1) + V_2 \sin(\omega_2 t + \phi_2)$$
$$= \text{Re}\,[jV_1\,e^{j\omega_1 t}\,e^{j\phi_1}] + \text{Re}\,[-jV_2\,e^{j\omega_2 t}\,e^{j\phi_2}] \qquad (2.60)$$

assuming the Q-factors of both tanks to be sufficiently high for all except the resonant-frequency voltages to be short-circuited. The current flowing out of C_c to the right is then

$$i_c = \frac{d}{dt}(C_c v_c) \qquad (2.61)$$

Although this current contains components at $(\omega_3 \pm \omega_1)$ and $(\omega_3 \pm \omega_2)$, because of the high Q-factor assumption, only the current component at $(\omega_2 - \omega_1) = \omega_2$ produces a voltage across tank 2, and similarly only that at $(\omega_3 - \omega_2) = \omega_1$ produces a voltage across tank 1. These two components of current are

$$i_c(\omega_1) = \frac{\omega_1 C_3}{2} V_2 \sin(\omega_1 t + \phi_3 - \phi_2)$$
$$= \text{Re}\left[-j\frac{\omega_1 C_3}{2} V_2\,e^{j\omega_1 t}\,e^{j(\phi_3-\phi_2)}\right] \qquad (2.62)$$

* The term *tank* indicates a circuit designed to store energy at a particular frequency.

Varactor Diodes

Similarly,

$$i_c(\omega_2) = \text{Re}\left[-j\frac{\omega_2 C_3}{2} V_1 e^{j\omega_2 t} e^{j(\phi_3 - \phi_1)}\right] \quad (2.63)$$

The effective admittance $Y(\omega_1)$, presented to the driving circuit is given by

$$Y(\omega_1) = \frac{\text{complex representation of } i_c(\omega_1)}{-\text{complex representation of } v(\omega_1)}$$

$$= -j\frac{(\omega_1 C_3/2) V_2 e^{j(\phi_3 - \phi_2)}}{jV_1 e^{j\phi_1}} = \frac{-\omega_1 C_3 V_2 e^{j(\phi_3 - \phi_2 - \phi_1)}}{2V_1} \quad (2.64)$$

If Y_2 is the admittance of tank 2, then

$$Y_2 = G_{T2} + j\left(\omega_2 C_2 - \frac{1}{\omega_2 L_2}\right) = \frac{\text{complex representation of } i_c(\omega_2)}{\text{complex representation of } v(\omega_2)} \quad (2.65)$$

From eqns. (2.60) and (2.63) we then have

$$V_2 = \frac{(\omega_2 C_3/2) V_1 e^{j(\phi_3 - \phi_2 - \phi_1)}}{Y_2} \quad (2.66)$$

Insertion of the complex conjugate of eqn. (2.66) into eqn. (2.64) gives the effective admittance presented to the driving circuit as

$$Y(\omega_1) = -\omega_1 \omega_2 \frac{C_3^2}{4} Y_2^* = -G \quad \text{(say)} \quad (2.67)$$

The power gain is taken as the ratio of power dissipated in the load conductance G_L to the available power from the generator (conductance G_g) and is given by

$$G_P = \frac{4G_g G_L}{(G_{T1} - G)^2} \quad (2.68)$$

G_{T1} is shown in Fig. 2.13.

The bandwidth between 3 dB points is found by setting the frequency deviation from resonance, δ, to that value which gives half the resonance gain and is found to be

$$2\delta = \frac{G_{T1} - G}{Q_1\left[G_{T1} + G\left(\frac{\omega_1 Q_2}{\omega_2 Q_1}\right)\right]} \quad (2.69)$$

Microwave Semiconductor Devices

Another item of importance is the gain-bandwidth product, written as

$$(\text{power gain})^{1/2} \times (\text{fractional bandwidth}) = \frac{2\omega_2(G_g G_L)^{1/2}}{\omega_2 Q_1 G_{T1} + \omega_1 Q_2 G}$$

$$\approx \frac{1}{Q_2} \frac{\omega_2}{\omega_1} \frac{2(G_g G_L)^{1/2}}{G} \quad (2.70)$$

provided the usual condition is satisfied that $\omega_1 Q_2 G \gg \omega_2 Q_1 G_{T1}$.

It may further be shown [11] that for a negative-resistance amplifier operated in conjunction with a circulator the gain-bandwidth product may also be written as

$$GB^{1/2} = 2\left(\frac{1}{B_S} + \frac{1}{B_I}\right)^{-1} \quad (2.71)$$

where B_S is the unpumped signal-circuit bandwidth and B_I is the unpumed idling-circuit bandwidth, which clearly shows that the bandwidth of neither circuit should be unnecessarily restricted and stray reactances should therefore be kept to a minimum. Maximization of operating bandwidths requires that the energies associated with the signal, pump and idler should each be confined to its appropriate region. In order to keep down the Q-factors, the signal and idling circuits are often of lumped form. It should be mentioned that travelling-wave-type parametric amplifiers using varactors have been made which have wide bandwidths, but these are complicated structures and are not in common use.

2.3.2 Harmonic Generators

Microwave harmonic generators are of two main forms. Both employ a transistor source which may give an output of several watts at a frequency between a few hundred megahertz and about 1 GHz. The first type employs a chain of low-order varactor multiplier stages which are amenable to analyses similar to those used for parametric amplifiers. The second type uses only a single variable-capacitance diode of the "step-recovery" type, whose fast switching action is used to impulse-excite an output resonator, giving frequency multiplication which may be 10 or even 20 times. The latter type of generator has tended to have smaller power capability, but there is evidence that such devices are no longer so limited.

Varactor Diodes

With non-linear resistors it has been shown by Page [13] that the maximum harmonic generation efficiency is restricted to

$$P_n \leqslant \frac{1}{n^2} P_1 \tag{2.72}$$

where P_1 and P_n are the available powers at the fundamental and nth harmonic frequencies. With varactors, however, using the Manley–Rowe equations, it can be shown that theoretical efficiencies may approach 100%. From these we note that, for a non-linear reactance with power flow at only f_1 and nf_1,

$$\frac{P_1}{f_1} + \frac{nP_n}{nf_1} = 0$$

Thus $P_n/P_1 = -1$, and if P_1 represents the power input to the generator, a conversion efficiency of unity is thereby indicated. This ideal is never

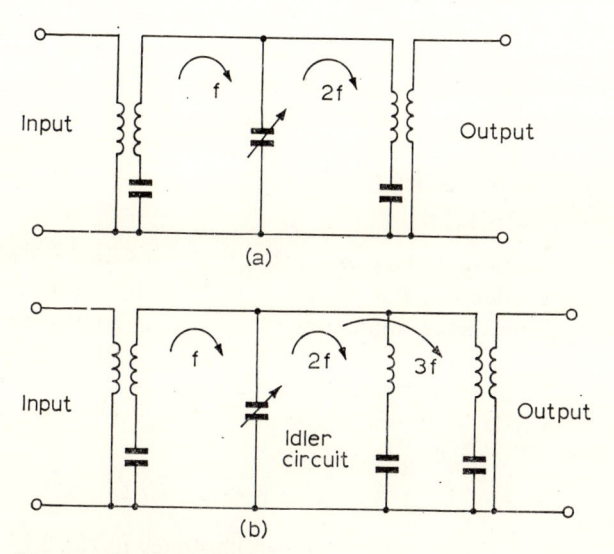

FIG 2.14 Multiplier stages
    (a) Basic (b) With idler

attained in practice because of losses associated with the circuit and with the series resistance of the varactor. The simplest type of circuit used in a multiplying chain is shown in Fig. 2.14(a), where the input tank is tuned to f_1 and the output tank to nf_1.

75

Chain Multipliers

A chain-type multiplier may employ up to five stages in getting from the transistor frequency to say 10 GHz. Each stage will usually multiply by either 2, 3, 4 or 6 times. For multiplication of higher order than two, an idler circuit at one or more of the intermediate harmonic frequencies is usually employed, as this can greatly improve the efficiency. For instance, with a tripler, an idler circuit tuned to the second harmonic would be employed, as indicated in Fig. 2.14(b). Many other forms of circuit are, of course, possible.

The efficiency of a frequency multiplier using a shunt arrangement (i.e. voltage excited) may be expressed approximately as [14]

$$\eta = 1 - \frac{kf_1}{f_c(V_B)} \tag{2.73}$$

where f_1 is the input frequency, $f_c(V_B)$ is the cut-off frequency of the varactor corresponding to the breakdown voltage, and is a constant which depends on the order of harmonic generation. The factor k is approximately 20 for a doubler and 35 for a tripler [15].

The power-handling capacity for maximum efficiency is given by

$$P_{in} = \frac{1}{A} \omega C_{min} (\phi + V_B)^2 \tag{2.74}$$

where V_B is the breakdown voltage, ϕ is the voltage above zero at which the diode becomes conducting, and A is a constant which depends on the order of multiplication, being approximately 35 for a doubler and 42 for a tripler.

In cascading a chain of multipliers many practical problems have to be solved which include hysteresis caused by dynamic detuning, parametric oscillation and relaxation oscillations.

Impulse Generators

The principle of action of impulse generators is illustrated in Fig. 2.15 [16]. Essentially we have a voltage source, drive inductance, step-recovery diode, bias battery and load. Separate equivalent circuits may be ascribed to the diode according to whether it is in the low- or high-impedance state. Conditions during the conduction interval are shown at (a), those during the depletion interval at (b), those during the second complete cycle at (c).

Varactor Diodes

By replacing the resistive output load of the impulse generator with a resonant tank circuit, a new spectrum at the output will be obtained, containing the same total energy but concentrated around the *n*th harmonic, to which the output circuit is tuned. The amount of energy in the *n*th spectral line is about two-thirds of the total, and consequently highly

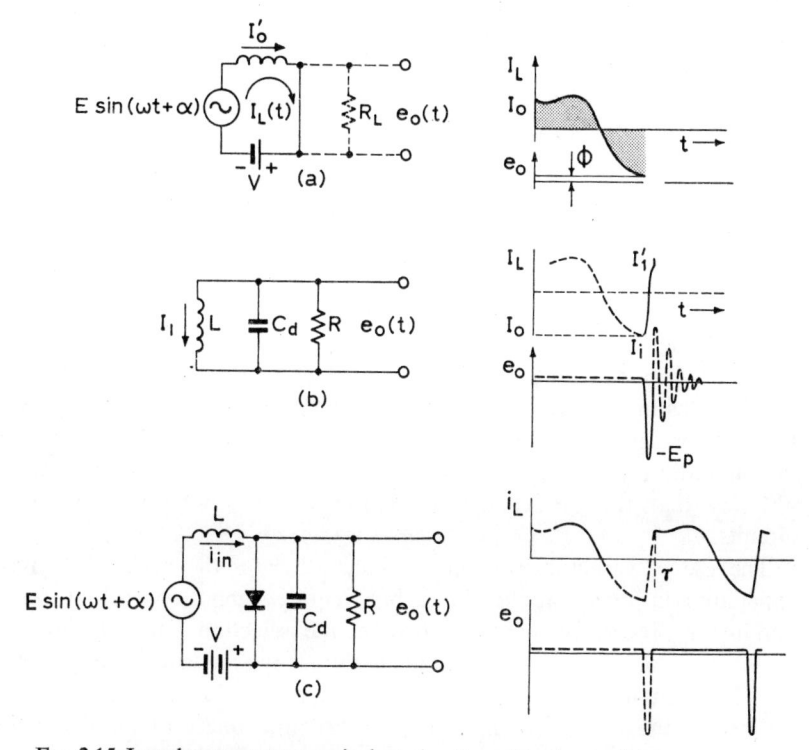

FIG 2.15 Impulse generator: equivalent circuits and basic conditions
    (*a*) During conduction interval
    (*b*) During impulse interval
    (*c*) During second complete cycle
    (*From Ref. 16*)

efficient harmonic generation is possible using that energy alone. Further filtering may be done to reduce the sideband level to the desired degree.

2.3.3 Varactor Switching of Microwave Signals
Varactors are used in switching applications, particularly where high speed is required without high power capability. Their use is in general,

however, restricted to frequencies below the microwave range on account of a limitation in switching ratio, as discussed below.

Consider the varactor as a shunt element across a transmission line. The high-impedance condition is achieved by tuning out the capacitance, the diode series resistance r then appearing as an effective parallel resistance $Q^2 r$. In the low-impedance state, the capacitance will become very large and the impedance will approach that of the series resistance alone. Neglecting, to a first order, any change in diode series resistance between the two states, the switching ratio will therefore be equal to Q^2. Now at, say, 10 GHz a relatively high-grade varactor having $f_c = 100$ GHz will have a Q-factor of only 10, with a corresponding switching ratio of 100. This compares very unfavourably with the switching ratios of several thousand easily obtainable with p–i–n diodes, discussed in Chapter 6, which may be made with high power capability also.

2.3.4 Tuning Varactors

Varactors are replacing mechanically tuned capacitors in a variety of applications in the frequency range from 1 MHz to about 1 GHz. Capacitance ratios between extremes of permissible bias of up to 20 or so may be used in radio/television, but tuning ratios in terms of frequency are appreciably less than this on account of fixed capacitance present in the tuned circuits.

The use of varactors tracking together obviates the need for ganged capacitors in many applications, but requires good control of semiconductor processing and may involve the selection and matching of diodes with nearly identical characteristics. However, since the Q-factor for a given cut-off frequency decreases in proportion to the operating frequency, the use of tuning varactors becomes uneconomic above the lower end of the microwave spectrum.

2.3.5 Limiter Varactors

This type of varactor is used as additional protection, after a gas T-R cell, for receivers employing tunnel-diode amplifiers or crystals which are particularly sensitive to burnout. The device must be fast-switching and the optimum geometry is substantially the same as for the step-recovery diode. Although the leakage "spike" energy from the T-R cell is relatively high (up to $0 \cdot 3 \times 10^{-7}$ J per pulse), the limiter varactor is required to absorb power for only a very short time, so that it can be physically very

Varactor Diodes

small. The breakdown voltage need not be high, but cut-off frequency at zero bias must approach 100 GHz for useful operation at a frequency of 10 GHz.

2.4 MANUFACTURE

Manufacturing techniques vary considerably according to the application, the manufacturer and the "state of the art". We here quote as examples details of constructions used for silicon varactors [17] and for a range of gallium-arsenide devices covering three frequency bands [3]. These particular varactors are intended for use in parametric amplifiers and also in frequency multipliers.

2.4.1 Silicon Varactors

These diodes employ a p^+–n–n^+ structure produced through diffusion techniques. Phosphorus is diffused into both sides of a 50 mΩ-cm slice of n-type silicon to form a 40 μm thick n^+-layer, which is subsequently lapped off on one side. Boron is then diffused into the lapped side to a depth of 20 μm to form the p–n junction, the boron surface concentration being 10^{20} atoms/cm³.

The n-region is then about 10 μm thick. Each slice is subsequently nickel plated on both sides, tinned on the n^+-side and cut into 0·8 mm squares. The dice are then mounted on tinned-nickel pans to form a low-resistance contact to the n^+-layer, and a mesa is formed by ultrasonic cutting to a depth of 35 μm into the p-side. After etching, the units are pushed into cans so that the top of the mesa makes a pressure contact to the can centre pin, such a structure being suitable for direct insertion into coaxial and waveguide circuits.

2.4.2 Gallium Arsenide Varactors

10 cm-band Diode

For this band Foxell and Wilson [3] have described a diffusion process in conjunction with a planar structure in epitaxial material. By employing a zinc diffusion to obtain a junction in n-type gallium arsenide a greater capacitance/voltage sensitivity is achieved, since unlike the majority of diffusants in other semiconductors which result in an error-function profile (see Chapter 3), this particular combination is anomalous and approaches the abrupt profile given by an alloyed junction.

Microwave Semiconductor Devices

The breakdown voltage with the types of junction described here is found empirically to be given by

$$V_B = 250\rho_n^{0.6} \tag{2.75}$$

over the range from 5 to 100 V. This expression is used to give the lowest possible resistivity compatible with the required breakdown voltage (6 V) so as to minimize the bulk resistance. The slices are chemically polished to a flat surface, and zinc is diffused in under an excess pressure of arsenic.

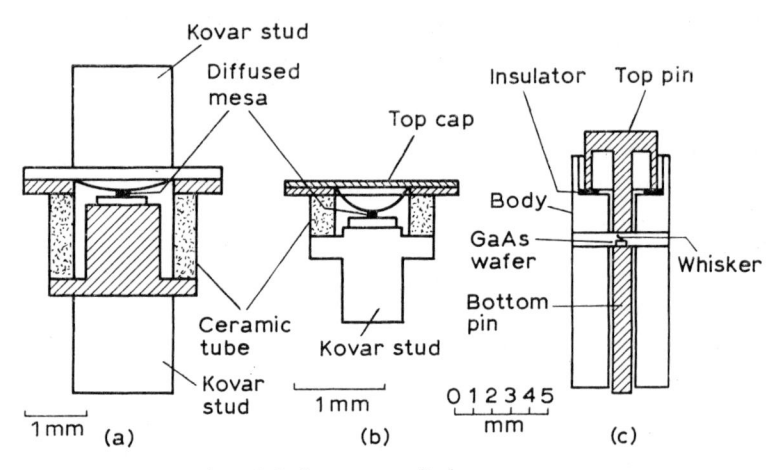

FIG 2.16 Construction of GaAs varactor diodes
(a) 10 cm band
(b) 3 cm band
(c) 1 cm band
(*From Ref.* 3)

The lattice of gallium arsenide is asymmetrical, and there is therefore an arsenic-rich face and a gallium-rich face to each slice cut from the ingot. The junction on the latter side is removed to retain the flatter junction on the arsenic-rich side, which assists in minimizing the p-type layer thickness and hence minimizes the resistance.

The slice containing a p–n junction is then diced into wafers and mounted in the encapsulation shown in Fig. 2.16(*a*). Gold wires are attached to a metal contact on the top of the wafer and to the flange, the junction characteristics being monitored whilst the junction size is reduced by etching to give a capacitance of about 0·5 pF.

3 cm-band Diode

Compared with the previous example, improvements are necessary in both diode element and encapsulation for use at a wavelength of 3 cm. An epitaxial structure is used in which a thin film of high-resistivity *n*-type gallium arsenide is grown on to a heavily doped substrate. The junction is formed in this layer, and thus the breakdown voltage is that associated with high-resistivity material, whilst the low-resistivity substrate reduces the bulk resistance.

An epitaxial layer of about $10\,\mu$m is used, and the junction depth is controlled to within $1\,\mu$m. A mesa is formed as before, but the wafer is mounted in a much smaller micropill case, illustrated in Fig. 2.16(*b*), which reduces the series inductance from about 640 pH to about 100 pH and raises the series resonant frequency from about 10 GHz to 30 GHz.

1 cm-band Diode

Still higher performance is required from both the semiconductor element and the encapsulation to permit varactor operation at 1 cm wavelength. Pending an advance in ceramic technology which would enable a suitable micropill case to be made, experimental units have used a point-contact structure with the diode fabricated directly into waveguide, as illustrated in Fig. 2.16(*c*). The body of the diode contains a section of reduced-height waveguide into which are inserted two pins, one carrying the semiconductor wafer and the other carrying the whisker, which also forms the choke section. The junction is formed by advancing the whisker until contact is made with the gallium arsenide, and then low voltage pulses are applied whilst the microwave-transmission characteristics are monitored continuously on an oscilloscope. The junction is produced by diffusion of *p*-type impurity from the whisker, usually copper or zinc, into the *n*-type gallium arsenide, thereby giving a shallow junction of small area.

2.4.3 Fabrication of Other Types

Other variable-capacitance devices are fabricated by techniques which are basically similar to those quoted above, the details being largely governed by the application. With tuning varactors, for instance, a well-defined law of capacitance versus voltage variation is desirable over a wide capacitance range, and this is best obtained by forming the junction in a relatively thick high-resistivity epitaxial layer of uniform composition. "Steprecovery" diodes, on the other hand, require a high and uniform "built-in" electric field to assist in carrier removal. This is achieved by having high-resistivity material in the vicinity of the junction with a rapid transition into *p*- and *n*-type regions, respectively, on either side. To this end it is possible to use a heavy double diffusion into high-resistivity material or alternatively to employ a single diffusion into an epitaxial structure containing a high-resistivity layer having an abrupt transition with the substrate.

2.5 VARACTOR MEASUREMENTS

The principal measurements are aimed at determining the values of elements in the equivalent circuit of Fig. 2.2, from which f_c may be obtained. Also, for parametric applications, the capacitance non-linearity factor is required.

It is usual to measure total capacitance by a bridge method (e.g. transformer ratio-arm type) at a frequency of about 1 MHz, where the series inductance and resistance are unimportant. A method of separating the junction capacitance from the stray capacitance is as follows [18].

The total capacitance is measured at zero bias (C_0) and at one other bias voltage (C_b). For each diode C_b is plotted against the difference $C_0 - C_b$. The resulting plot is substantially a straight line from whose slope an approximate power law of capacitance versus voltage variation may be deduced. The intercept on the axis of C_b gives directly the stray capacitance C_p.

Microwave measurements are needed for establishing the other circuit parameters. Use may be made of slotted-line impedance measurements for this purpose, but since v.s.w.r.s are very low, it is difficult to obtain accurate information on the effective resistance. A more satisfactory method which provides all the equivalent circuit information, due to DeLoach [19], involves the measurement of transmission past a diode mounted between the broad faces of a reduced-height waveguide. We here describe the method given by Roberts and Wilson [18], which is basically that of DeLoach but does not involve the use of a swept-frequency source. The technique is applicable to encapsulated diodes if the package capacitance has a reactance which is large compared to the diode series resistance at the series resonant frequency, given by

$$f_s = \frac{1}{2\sqrt{(L_s C_j)}} \qquad (2.76)$$

Varactor Diodes

The mounting of the diode is illustrated in Fig. 2.17, the taper sections being well matched. The waveguide size is chosen to be appropriate to the frequency band containing f_s.

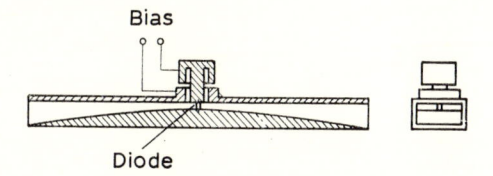

FIG. 2.17 Waveguide holder used for transmission measurements near the series resonant frequency
(*From Ref.* 18)

At frequencies near to f_s it is usually found that the reactance of the package capacitance and the detector impedance are sufficiently high to have negligible shunting effect, so that the equivalent circuit reduces to a series combination of L_s, r and C_j shunting the transmission line, and the transmission loss can be measured with the arrangement of Fig. 2.18. The following procedure may then be used.

(*a*) The frequency is found which gives minimum transmission, which occurs at series resonance. The equivalent circuit is then simply the diode resistance r, which is given by

$$r = \frac{Z_0}{2(\sqrt{(T)} - 1)} \qquad (2.77)$$

where T is the measured transmission loss, and Z_0 is the characteristic impedance of the reduced-weight waveguide, using the power-voltage definition of impedance.

(*b*) The variation of resonant frequency with bias may be used to obtain the capacitance law, since

$$C_j = \frac{1}{4\pi^2 L_s f_s^2} = \frac{1}{K f_s^2} \qquad (2.78)$$

Plotting KC_j ($=1/f_s^2$) against voltage bias on a logarithmic scale will thus give a straight line for a diode whose capacitance versus bias relationship obeys a simple power law, the slope of the line giving the exponent.

(*c*) With the frequency set to f_s the forward bias V_1 and reverse bias V_2 required to double the transmitted power are measured. From the plot

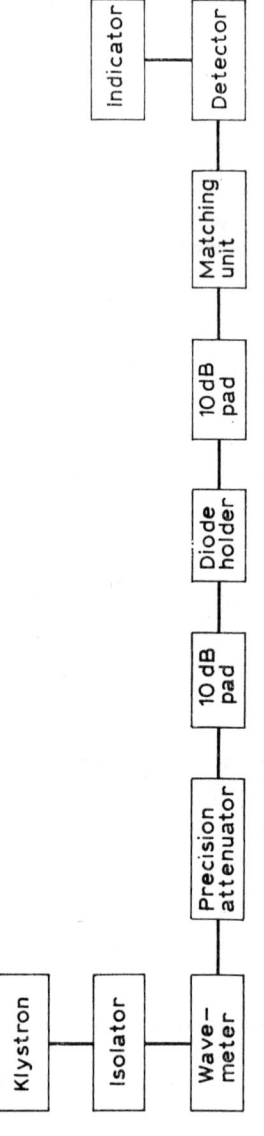

Fig 2.18 Apparatus for transmission measurements near the series resonant frequency
(From Ref. 18)

Varactor Diodes

of $1/f_s^2$ versus bias, corresponding values of KC_{j1} and KC_{j2} can be found, the Q-factor of the diode then being given by

$$Q = \frac{KC_{j1} + KC_{j2}}{KC_{j1} - KC_{j2}} \tag{2.79}$$

(*d*) The cut-off frequency of the diode at zero bias is thus

$$f_{c(0)} = Qf_{s(0)} \tag{2.80}$$

(*e*) The junction capacitance at zero bias is thus

$$C_{j(0)} = \frac{1}{2\pi r f_{c(0)}} \tag{2.81}$$

which enables the plot of KC_j versus bias to be calibrated.

(*f*) The series inductance is given by

$$L_s = \frac{1}{4\pi^2 f_{s(0)}^2 C_{j(0)}} \tag{2.82}$$

(*g*) Taking the applied bias voltage for a forward current of $1\,\mu\text{A}$ as V_F, values of $KC_{j(V_F)}$ and $KC_{j(-1)}$ may be read off the curve of KC_j versus bias and the capacitance non-linearity factor determined:

$$\frac{KC_{j(V_F)} - KC_{j(-1)}}{2(KC_{j(V_F)} + KC_{j(-1)})} \tag{2.83}$$

Roberts and Wilson also described a method of measuring the diode parallel self-capacitance C_p, in which the diode is mounted in the inner of a very short piece of coaxial line connecting two waveguide-to-coaxial transitions. This permits a measurement of the parallel resonant frequency f_p by finding the minimum of transmission with the diode mounted in series with the source and a load, without the necessity of using coaxial connectors. (These present serious difficulties at the frequency band in which f_p usually occurs, i.e. around 30 GHz.) Using this technique, C_p is then obtained from the relationship

$$C_p = \frac{C_j}{\left(\dfrac{f_p}{f_s}\right)^2 - 1} \tag{2.84}$$

The junction capacitance measured at microwave frequencies is found to be higher than that measured at 1 MHz by a fixed value of about 0·1 pF which is independent of the bias voltage. This effect is explained by assuming the modified equivalent circuit of Fig. 2.2(b). It can be shown that, if $\omega C_i' r \ll 1$,

$$r \approx \frac{r'}{\left(\frac{C_i'}{C_j'} + 1\right)^2}$$

and

$$C_j \approx C_j' + C_i' \tag{2.85}$$

These equations explain observations that R_s is a function of bias and also that the junction capacitance measured at microwave frequencies is higher than the true junction capacitance by an amount C_i'.

2.5.1 Other Methods for Measuring Q-factor

Houlding [20] first suggested a method of Q-factor measurement based on a simple series equivalent circuit which has been widely used in the 10 cm band. In this method the diode is initially matched into the measuring waveguide system with some arbitrary bias applied (usually zero). The bias voltage is then changed, thereby producing a change ΔC in the junction capacitance, the Q-factor then being given by

$$Q_0 = \frac{C_0 + \Delta C}{\Delta C} \Delta Q \tag{2.86}$$

where ΔQ is the normalized change in reactance, and C_0 is the effective series capacitance in the matched condition.

A somewhat similar method was proposed by Harrison [21] which permits the direct determination of Q_0 if the semiconductor region can be replaced by an effective short-circuit. This can sometimes be achieved by passing a large forward current through the varactor, but it is not always possible to achieve an effective short-circuit at currents which do not load the diode excessively, particularly with gallium arsenide. Mavaddat [22] has given an extension to Harrison's method which obviates the necessity of producing an effective short-circuit and permits the determination of both Q_0 and C_0, as described below.

If $+jX$ is the reactance of the diode capacitance normalized to the effective series resistance and $-jB$ is the corresponding normalized susceptance, such that $jB = 1/-jX$, then obviously

$$XB = 1 \tag{2.87}$$

Varactor Diodes

For the original bias value, i.e. with the diode matched to the line, $C = C_0$ and

$$X_0 B_0 = 1 \tag{2.88}$$

Let ΔX and ΔB represent changes in reactance and susceptance, respectively, corresponding to a change in effective capacitance ΔC; then eqn. (2.87) may be written as

$$(X_0 + \Delta X)(B_0 + \Delta B) = 1 \tag{2.89}$$

From eqns. (2.88) and (2.89),

$$\frac{X_0}{\Delta X} + \frac{B_0}{\Delta B} = -1$$

and hence

$$\frac{X_0}{\Delta X} + \frac{C_0}{\Delta C} = -1 \tag{2.90}$$

A plot of $1/\Delta X$ versus $1/\Delta C$ is therefore a straight line which intersects the $1/\Delta X$ axis at $1/X_0$ and the $1/\Delta C$ axis at $1/C_0$. The reciprocals of these two intercepts are Q_0 and C_0, respectively, where $Q_0 = 1/\omega C_0 r$.

Other methods of measuring Q-factor have also been used which involve the use of calibrated impedances.

The author has proposed an alternative procedure which is based on the method due to Houlding but avoids both the necessity for matching and the use of impedance standards [23]. It also lends itself to a procedure in which Q-factor may be indicated directly with the aid of an automatic Smith diagram display unit of a type which can be used with power levels at the load down to the order of a microwatt [24].

Let C_1 and C_2 be the values of effective series capacitance corresponding to two settings of bias voltage. For fixed values of these voltages the ratio C_1/C_2 will have a constant value K for every diode, which can be found from the capacitance versus voltage relationship. Since for a given series resistance the Q-factor is inversely proportional to capacitance, if Q_1 and Q_2 are values corresponding to C_1 and C_2 respectively, then

$$Q_2/Q_1 = K \tag{2.91}$$

The difference in Q-factor between the two bias conditions is thus given by

$$\Delta Q = Q_1 - Q_2 = Q_1(1 - K) \tag{2.92}$$

By choosing appropriate bias voltages, K may be made any value we please. A convenient value is 0·9, giving $Q_1 = 10 \Delta Q$. The determination of ΔQ without matching is explained below.

For a lossless transformation between two impedance planes Z_1 and Z_2, which correspond in the sense that $Z_2 = \infty$ when $Z_1 = \infty$, it may be shown that

$$Z_2 = \alpha Z_1 + j\beta \tag{2.93}$$

where the constants α, β are real numbers [25].

Consider two values of impedance at the first plane, Z_1' and Z_1'', which differ only in their reactive components, i.e.

$$Z_1' = R + jX$$
$$Z_1'' = R + j(X + \Delta X)$$

Corresponding impedances, Z_2' and Z_2'', at the second plane are given by

$$Z_2' = R_2' + jX_2' = \alpha R + j(\alpha X + \beta)$$
$$Z_2'' = R_2'' + jX_2'' = \alpha R + j[\alpha(X + \Delta X) + \beta]$$

Hence

$$\frac{X_2'' - X_2'}{R_2''} = \frac{\Delta X}{R} = \Delta Q \tag{2.94}$$

Thus, where a normalized reactance change at the first plane is interpreted as a change in Q-factor, the change in normalized reactance at a second plane, which corresponds in the sense described, will indicate the same value ΔQ independently of the transformation constants α, β.

It is implicit in eqn. (2.93) that circles of constant resistance on a Smith diagram transform at a corresponding plane into the same family of circles at the other, but for each specific circle the resistance value is changed in the ratio $\alpha:1$. Thus, if the first plane Z_1 refers to the varactor junction itself, a change in capacitance alone at that plane will vary the impedance at a corresponding measuring plane Z_2 along a circle of constant resistance at the second plane also. Thus, to find such a corresponding plane from an arbitrarily chosen measuring plane, it is only necessary to rotate the

Varactor Diodes

Smith diagram appropriate to the arbitrary plane until the impedance locus produced by a change in capacitance at the junction lies along a circle of constant resistance, when the reactance change at plane Z_2, normalized to the resistance, will indicate ΔQ.

The procedure is facilitated by introducing a grid of lines for fixed values of Q-factor on the Smith diagram appropriate to the plane Z_2. This grid is determined as described below.

The normalized impedance Z for any measuring plane is given in terms of the complex reflection coefficient ρ by

$$Z = R + jX = \frac{1+\rho}{1-\rho} \tag{2.95}$$

Hence

$$R = \frac{1 - \rho\rho^*}{1 + \rho\rho^* - \rho - \rho^*}$$

$$jX = \frac{\rho - \rho^*}{1 + \rho\rho^* - \rho - \rho^*}$$

Thus

$$\frac{jX}{R} = \frac{\rho - \rho^*}{1 - \rho\rho^*} = jQ$$

whence

$$\rho\rho^* + \frac{j}{Q}\rho^* - \frac{j}{Q}\rho - 1 = 0 \tag{2.96}$$

Equation (2.96) is seen to represent the equation of a circle on comparing it with the general circle equation

$$(\rho - a)(\rho^* - a^*) = K^2 \tag{2.97}$$

with the radius K given by

$$K = \left(1 + \frac{1}{Q^2}\right)^{1/2} \tag{2.98}$$

and the centre displaced from the origin of co-ordinates by a vector $a = j/Q$.

A family of circles representing constant Q may thus be constructed on a Smith chart, and these, together with the family of constant-resistance

Microwave Semiconductor Devices

circles, suffice to determine ΔQ. The resulting new chart is shown in Fig. 2.19. To obtain the cut-off frequency f_c, the diagram is orientated so that the measured normalized-impedance points corresponding to the two bias conditions both fall on the same constant-resistance circle. The corresponding Q values are then read off the chart, and assuming bias voltages chosen to give $Q = 10\,\Delta Q$, the cut-off frequency is given by $f_c = 10\,f\Delta Q$, where f is the measuring frequency. By the use of an automatic Smith diagram display unit of a type which can be used down to

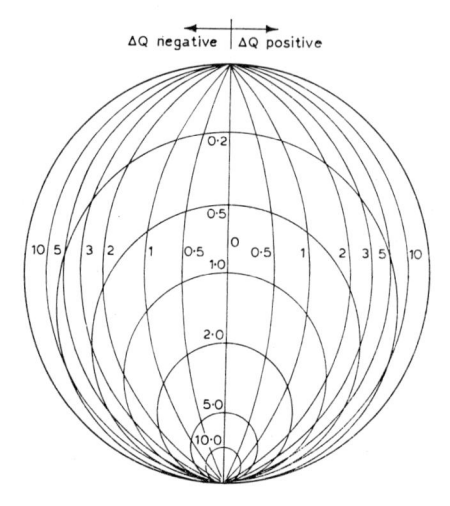

FIG 2.19 Modified Smith chart (constant Q and resistance circles)
(*From Ref.* 23)

microwatt power levels incident on the varactor, a direct reading of ΔQ or of f_c may be obtained.

It is to be noted that the method described above assumes a simplified series circuit to be an adequate representation of the varactor, with the resistance unchanged between the two bias states. This assumption is also implicit in the methods of Houlding, Harrison and Mavaddat.

Other methods of measuring Q-factor have been considered, including the use of a cavity resonator. However, these generally demand careful setting up as well as calibrations by impedance standards and do not appear to have found widespread favour.

Varactor Diodes

REFERENCES

1 Howson, D. P., Owen, B., and Wright, G. T., "The space-charge varactor", *Solid State Electron.*, **8**, p. 913 (1965).

2 Uhlir, A., Jr., "The potential of semiconductor diodes in high-frequency communications", *Proc. Inst. Radio Engrs*, **46**, p. 1099 (1958).

3 Foxell, C. A. P., and Wilson, K., "Gallium arsenide varactor diodes", *Radio Electron. Engr*, **31**, p. 245 (1966).

4 Jonscher, A. K., *Principles of Semiconductor Device Operation* (Bell, 1960).

5 Shurmer, H. V., "Intrinsic frequency limitations for semiconductor microwave devices", *Radio Electron. Engr*, **31**, p. 93 (1966).

6 Moll, J. L., Krakauer, S., and Shen, R., "P–N junction charge storage diodes", *Proc. Inst. Radio Engrs*, **50**, p. 43 (1962).

7 Hilibrand, J., Mueller, C. W., Stocker, C. F., and Gold, R. D., "Semiconductor parametric diodes in microwave computers", *Trans. Inst. Radio Engrs*, **EC-8**, p. 287 (1959).

8 Manley, J. M., and Rowe, H. E., "Some general properties of non-linear elements: Part 1—general energy relations", *Proc. Inst. Radio Engrs*, **44**, p. 904 (1956).

9 Weiss, M. T., "Quantum derivation of energy relations analogous to those for non-linear reactances", *Proc. Inst. Radio Engrs*, **45**, p. 1012 (1957).

10 Aitchison, C. S., Humphreys, B. L. and Neufeld, E. L., "The overall noise figures of diode parametric amplifier systems", *Proc. Inst Elect. Engrs*, **110**, p. 348 (1963).

11 Aitchison, C. S., Davies, R., and Gibson, P. J., "A simple diode parametric amplifier design for use at S, C and X-band", *Proc. Joint Symposium on Microwave Applications of Semiconductors*, University College, London, Paper 26 (1965).

12 Heffner, H., and Wade, G., "Gain, bandwidth and noise characteristics of the variable-parametric amplifier", *J. Appl. Phys.*, **29**, p. 1321 (1958).

13 Page, C. H., "Harmonic generation with ideal rectifiers", *Proc. Inst. Radio Engrs*, **46**, p. 1738 (1958).

14 Penfield, P., and Rafuse, R., *Varactor Applications* (MIT Press, Cambridge, Mass., 1962).

15 Edwards, B., and Madan, A., "An X-band solid-state source using frequency multipliers", *Proc. Joint Symposium on Microwave Applications of Semiconductors*, University College, London, Paper 34 (1965).

16 Hamilton, S., and Hall, R., "Shunt-mode harmonic generation using step-recovery diodes", *Microwave J.*, **10**, No. 5, p. 69 (1967).

17 Ingless, R., "Variable Capacitor Diodes", *GEC. J.*, **29**, p. 80 (1962).

18 Roberts, D. A. E., and Wilson, K., "Evaluation of high-quality varactor diodes", *Radio Electron. Engr*, **31**, p. 277 (1966).

19 DE LOACH, B. C., "A new microwave measurement technique to characterize diodes and an 800 Gc/s cut-off frequency varactor at zero volts bias", *Trans. IEEE*, **MTT-12**, p. 15 (1964).
20 HOULDING, N., "Measurement of varactor quality", *Microwave J.*, **3**, p. 40 (1960).
21 HARRISON, R. I., "Parametric diode Q-measurements", *ibid.*, **3**, p. 43 (1960).
22 MAVADDAT, R., "Varactor diode Q-factor measurements", *J. Electron. Contr.*, **15**, No. 1, p. 51 (1963).
23 SHURMER, H. V., "Microwave measurement of varactor diode Q-factors without matching or standard impedances", *ibid.*, **17**, No. 5, p. 491 (1964).
24 SHURMER, H. V., "An automatic smith diagram display unit for use at low power levels", *Proc. Instn Elect. Engrs*, **104B**, p. 507 (1957).
25 SHURMER, H. V., "Transformations of the Smith chart through lossless junctions", *ibid.*, **105C**, p. 177 (1958).
26 DANIELSON, W. E., *J. Appl. Phys.*, **30**, p. 8 (1959).

3

Schottky-Barrier Diodes

3.1 INTRODUCTION

A Schottky-type junction is one arising from the potential barrier between a semiconductor and a metal of substantially greater work function in which the following conditions are fulfilled:
1. The barrier height viewed from the semiconductor is large compared with kT/e.
2. The thickness of the depletion layer is large compared with an electron wavelength, so that tunnelling is negligible.
3. The concentration of both electrons and holes within the depletion layer is much less than that of ionized impurities.

A Schottky-barrier diode is essentially one having a junction such as described above on the one side of a semiconducting wafer and an ohmic contact on the opposite side.

In point-contact crystal diodes the rectifying element approximates to a Schottky-type structure, but the term *Schottky-barrier diode* has come to be identified with devices possessing junctions which are essentially different from the "formed" point-contact variety. In the diodes we now describe, the metal associated with the barrier is deposited on a clean semiconductor surface by processes such as evaporation, sputtering or electrochemical means, these devices being relative newcomers to the microwave field.

With Schottky-barrier diodes, the junction areas are considerably larger than for point-contact structures, but the capacitance is kept down to values which are of the same order, usually by using epitaxially prepared semiconductor material with films of high resistivity on substrates of low resistivity.

The junction areas are delineated photo-lithographically, diameters ranging up to about 25μm. It is customary to deposit the metal through holes prepared in an oxide film covering the semiconductor surface, this film being of the order of 1 µm in thickness, and the semiconductor epitaxial film a few tenths of a micron.

Compared with point-contact diodes, Schottky-barrier diodes have several advantages. They can have values of Cr product which are lower, resulting in improved noise figures with mixers, for example. They are also more rugged and reproducible, requiring none of the "forming" techniques used with point-contact devices. A wider range of impedance values is available and higher burnout ratings are claimed. However, it should be added, in connection with this last point, that although higher mean powers can undoubtedly be withstood, it is not at present certain that diodes with improved noise figures can also have better resistance to multiple pulses of a few nanoseconds' duration, and it is suggested that voltage breakdown may be a limitation here. Schottky-barrier diodes, by virtue of the robustness of their contacts, can also readily be applied to microwave integrated circuits. As with point-contact diodes, these devices also depend essentially on the action of majority carriers—there is very little charge storage and the response is very fast.

Silicon and gallium arsenide are the semiconductor materials in commercial use for Schottky-barrier diodes, the metallic side being typically gold, nickel or palladium, although many other metals have been used for a variety of specific applications. Gallium arsenide is superior to silicon as regards Cr product.

A consequence of making junctions in the ways indicated is that much lower reverse currents are obtained than with point-contact diodes (nanoamperes rather than microamperes at -1 V). This leads to mixers with noise ratios at conventional i.f. frequencies (10–60 GHz) close to unity, and to flicker noise performance which is superior to that of point-contact diodes. However, for detectors the same feature leads to excessive impedance values at zero bias, so that some forward bias must be employed.

At the higher microwave frequencies, conversion losses unattainable with point-contact mixers are possible, values in the range 5–6 dB at 55 GHz having been reported for gold/gallium-arsenide diodes [1]. As a consequence of the low Cr product the conversion loss and hence the noise figure do not become degraded by parasitic resistance and capacitance until frequencies well beyond 10 GHz, are reached, as illustrated in Fig. 3.1.

Diode manufacturers are aiming to use these devices as replacements for point-contact mixers, biased detectors and some varactors, as well as seeking new applications appropriate to the features indicated, e.g. fast switching.

Schottky-barrier Diodes

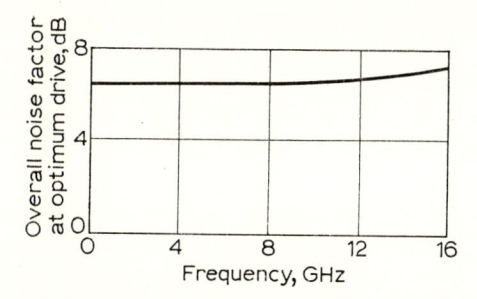

Fig 3.1 Overall noise factor v. frequency for typical GaAs Schottky-barrier diode
Noise figure of i.f. amplifier = 1·5 dB

3.2 THEORY

The basic theory of Schottky-barrier junctions has already been indicated in Chapter 1, leading to the I/V relationship of eqn. (1.4). It was noted that for point-contact diodes the multiplying factor in the voltage exponential was in practice about half the value e/kT predicted by elementary theory. A more sophisticated theory, allowing for image forces, shows that the above factor should be taken as e/nkT, where $n = 1\cdot06$. In fact, values for n of around 2·0 are measured with point-contact diodes, but values close to 1·06 are measured on Schottky-barrier diodes prepared under optimized processing conditions [2]. Introducing the factor n and substituting $(2kT/\pi m^*)^{1/2}$ for the average thermal velocity $\bar{u}$ in eqn. (1.4), we obtain the d.c. characteristic in a form commonly quoted for Schottky diodes:

$$J = ne\left(\frac{kT}{2\pi m^*}\right)^{1/2} e^{-e\phi_0/kT}(e^{eV/nkT} - 1) \tag{3.1}$$

If the processing conditions are not optimized, results are obtained which suggest that thin layers of insulating material are formed on the surface. These have the effect of altering the localized charge at the surface and could explain apparent variations in ϕ_0 by a factor of up to about 2. In addition, the presence of such layers would reduce the amount of applied voltage available across the rectifying barrier. They would not, however, prevent rectification, being probably no more than a few ångströms thick* and therefore permitting large tunnelling currents. It can easily be shown that, under low applied voltage conditions, such insulating layers would

---

* 1 Å = 10^{-10} m.

also have the effect of increasing the apparent value of n in eqn. (3.1), which in practice may rise to values as high as 3 for junctions formed on poorly prepared surfaces [3]. These results on Schottky-barrier diodes strongly suggest that the same effect is responsible for the relatively poor performance obtained from point-contact diodes in relation to simple theory.

We have noted that epitaxially grown semiconductor material is usually required for microwave Schottky-barrier devices, although in principle alternative means could be used [4]. Ideally, to minimize the Cr product, one would expect to require a geometry in which the edge of the space-charge depletion layer just reached through to the interface between the high-resistivity film and the low-resistivity substrate. Such a situation would suggest a series resistance due only to the substrate, say $1\,\Omega$ for a junction with a diameter of $25\,\mu$m and a substrate resistivity of $0.005\,\Omega$-cm.

There are two basic factors which prevent such low values of series resistance being achieved in practice. Firstly, with mixers and varactors, the depletion layer boundary moves to and fro during the driving cycle, so that any high-resistivity material uncovered during the forward part of the driving cycle will contribute to the effective series resistance. The higher the resistivity of the epitaxial film the larger will be this effect, so that it is necessary to compromise on film resistivity. Secondly there is the question as to how sharp a transition is possible between substrate and film, a problem which we discuss below.

Epitaxial films need to be grown at high temperatures and thus tend to become contaminated at the outset by dopant material which diffuses from the substrate. Further and often more serious contamination can occur during diode preparation, particularly with p–n junction devices, but this can be avoided in methods which are available for preparing Schottky-barrier junctions. As an example we will consider epitaxial silicon appropriate to Schottky-barrier diodes, which is commonly grown by the hydrogen reduction of silicon tetrachloride. It has been shown that the doping profile of such a layer may be characterized approximately by an expression of the form

$$n = \frac{n_0}{2}\,\text{erfc}\,\frac{x}{2\sqrt{(Dt)}} + n_0^*\,e^{-\psi x} + A(1 - e^{-\psi x}) \qquad (3.2)$$

The first term in this equation is due to the diffusion from the substrate of doping level n_0 for a time t with diffusion constant D. The second term represents the substrate contribution through the gaseous exchange of

Schottky-barrier Diodes

dopant between the solid and gas phase with parameters n_0^* and ψ [5]. The last term represents the gas phase contribution, initially zero, but tending to the value A for thick films.

Figure 3.2 shows the impurity concentration appropriate to the above type of process for a range of values of effective diffusion time and gas phase

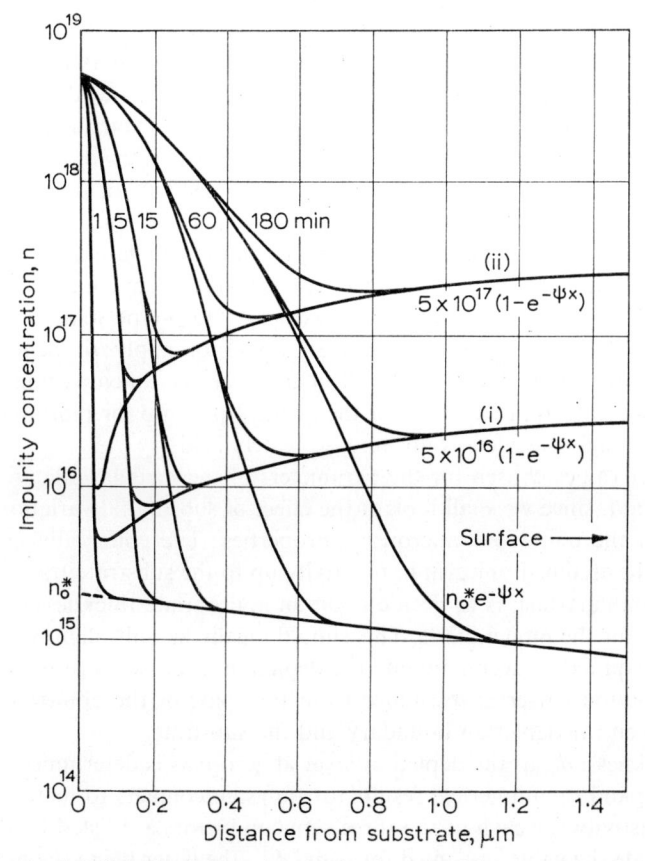

Fig 3.2 Calculated impurity profiles in epitaxial layers for selected diffusion times at 1 120°C
in eqn. (3.2),
$n_0 = 10^{19}$ atoms/cm^3
$D_{1\,120°C} = 3 \times 10^{-14}$ cm^2/s
$n_0^* = 2 \times 10^{15}$ atoms/cm
$\psi = 5.77 \times 10^3$/cm
$A_i = 5 \times 10^{16}$ atoms/cm^3
$A_{ii} = 5 \times 10^{17}$ atoms/cm^3

doping. Such impurity profiles are useful in enabling the Cr product and various other device parameters to be estimated for a particular junction, as we shall describe at the end of the following section.

3.3 DESIGN STUDY

Point-contact crystal diodes use semiconductor material with bulk resistivity of the order of $0{\cdot}01\,\Omega$-cm, although the surface resistivity is undoubtedly somewhat higher with silicon devices, owing to the out-diffusion of impurities resulting from heat treatments. However, as a starting point for a design study we shall assume model Schottky-barrier diodes using epitaxial semiconductor material with a film resistivity of $1\,\Omega$-cm, the semiconductors to be considered being those which have proved suitable for point-contact diodes, namely silicon, germanium and gallium arsenide, but n-type material will be assumed for all three semiconductors on account of its superior mobility. The restriction to p-type material with silicon, noted for point-contact devices, does not apply to Schottky-barrier diodes. The metal contact will in each case be taken as tungsten, which at one time appeared to offer the greatest promise for reproducible devices, although this view is not now so widely held.

The exact values chosen for the parameters of the model diode are not too important, since we shall look at the effect of substantial variations in these upon the computed microwave properties. The epitaxially grown layers will be assumed uniform in resistivity up to the substrate transition, and a preliminary task is to decide upon an appropriate thickness of epitaxial layer for the model diode. This should ideally be only slightly larger than that required to accommodate the depletion layer, so as to minimize the contribution to series resistance from that part of the epitaxial film lying between the depletion boundary and the substrate.

The thickness, d_0, of the depletion layer at zero bias is determined from eqn. (2.16), and the numerical results for tungsten contacts to material of $1\,\Omega$-cm resistivity for each of the three semiconductors are listed in Table 3.1, alongside the values assumed for n and ϕ_{ms}, the latter being the metal–semiconductor barrier potential [6]. It would appear from these values of d_0 that, for the model diodes, a suitable thickness w for the epitaxial layer would be $1\,\mu$m. The table also lists the capacitances corresponding to the quoted values of d_0 for junctions of $25\,\mu$m diameter and the corresponding series resistance due to the undepleted part of a $1\,\mu$m thick epitaxial layer plus the spreading resistance of substrate material of $0{\cdot}005\,\Omega$-cm resistivity,

Schottky-barrier Diodes

neglecting skin effect. It will be seen that these values of capacitance are of the same order as in point-contact microwave diodes. We will therefore regard it as satisfactory to assume for our model diodes a junction diameter of 25 μm together with a 1 μm thick epitaxial layer of resistivity 1 Ω-cm.

TABLE 3.1

	n (atoms/m³)	ϕ_{ms} (V)	d_0 (μm)	C (pF)	r (Ω)
Si	$\times 10^{21}$ 5·00	0·67	0·32	0·168	13·40
Ge	1·60	0·40	0·51	0·141	9·64
GaAs	1·14	0·71	0·76	0·074	4·84

We have first calculated the cut-off frequency f_c as that appropriate to varactor diodes, i.e. $1/2\pi Cr$. This has been done for a range of values of doping level and film thickness above and below those assumed for the model, and a comparison of the three semiconductor materials on this basis is shown in Fig. 3.3. It will be noted that the results depend most critically on film thickness. When this is reduced to a value corresponding to the thickness of the depletion layer, f_c tends towards very high values. With thicker films, however, the resulting increase in resistance quickly reduces f_c to very low values.

All of the curves of f_c versus doping level show a marked minimum. At the high doping end of the scale f_c rises, because the decrease in series resistance is more rapid than the increase in capacitance. For low doping levels, f_c increases because the associated increase in depletion layer thickness leaves progressively less undepleted high-resistivity material to contribute to the series resistance, which more than compensates for the increase in resistivity. These curves strongly indicate gallium arsenide as having the most suitable properties from the point of view of high-frequency parasitic degradation, germanium being indicated as second best and silicon third. However, there are other consideration which tend to make the choice of material rather less obvious, such as basic reproducibility problems with gallium arsenide and a tendency for high reverse leakage currents with germanium.

Microwave Semiconductor Devices

In Fig. 3.4 are shown the results of extending the treatment to the tangential sensitivity of biased detectors. The same model diode was used for normalizing purposes, the bias current density being taken as that giving a video resistance of 1 000 Ω for a junction of 25 μm diameter. The resistance and capacitance were calculated as before, allowing for the effect of the bias voltage on depletion layer thickness, and the tangential sensitivity was computed for a frequency of 9·375 GHz, using eqns. (1.9)

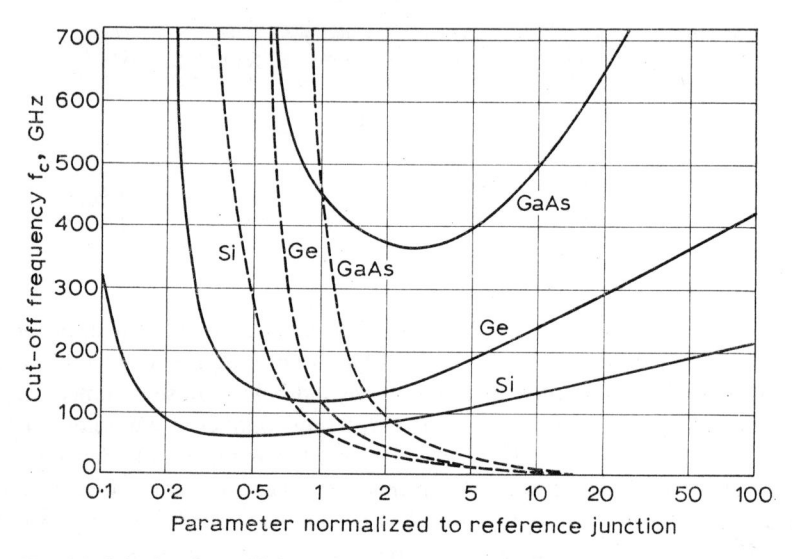

FIG 3.3 Calculated cut-off frequency v. normalized doping level and film thickness
Reference junction, 25 μm dia.; epitaxial layer, 1 μm; resistivity, 1 Ω-cm
——— doping level   - - - - film thickness

and (1.12). The noise resistance of the video amplifier was taken as 500 Ω and a bandwidth of 1 MHz was assumed.

The effect of varying the area is also shown in Fig. 3.4, in addition to doping level and film thickness. It was not appropriate to do this for the previous chart, since f_c is determined by the product Cr and in an ideal epitaxial situation this is substantially independent of area, whereas the corresponding factor in relation to tangential sensitivity is approximately C^2Rr, which is strongly dependent on area.

Figure 3.4 clearly places the three chosen semiconductor materials in the same order of merit as before. By decreasing any one of the parameters, however, it would appear possible to improve the tangential sensitivity

100

Schottky-barrier Diodes

up to a theoretical low-frequency value of 62 dBm in each case, but the relative difficulty involved in approaching the theoretical limit is linked to the order of merit for the materials, and it would clearly be unsatisfactory to seek high sensitivity by an arrangement which demanded too critical a control of any one variable.

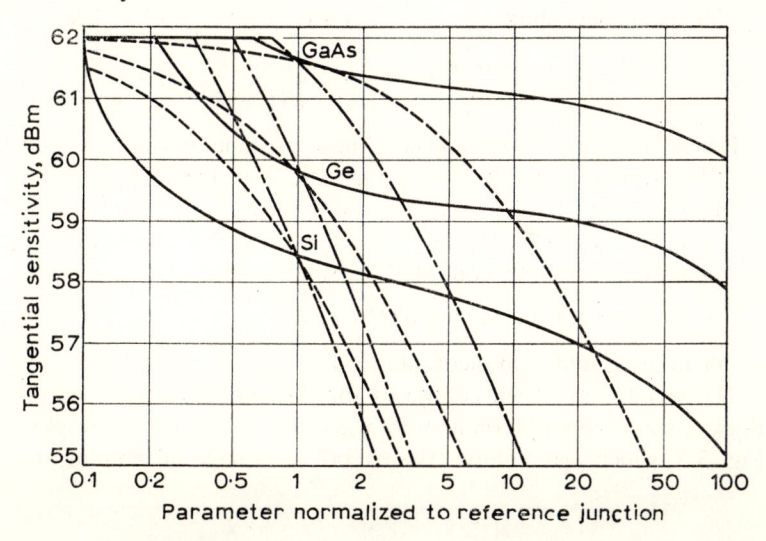

Fig 3.4 Calculated tangential sensitivities for selected values of doping, thickness and area
Reference junction, 25 μm dia.; epitaxial layer, 1 μm, resistivity 1 Ω-cm
——— doping level   —·—· film thickness   - - - - junction area

From the Cr product we have calculated the optimized conversion loss as a mixer, using a method due to Baron [7]. In this method, the mixer is represented as a "black box" of intrinsic loss as derived by Torrey and Whitmer, e.g. L_0 for the broadband mixer, as given by eqn. (1.29) [8]. The signal terminals of the box are assumed to be shunted directly by the barrier capacitance C, the spreading resistance r being placed in series with both the signal and i.f. terminals. It may then be shown that the microwave conversion loss is given by

$$L_0' = L_0 \left(1 + G_1 r + \frac{B^2 r}{G_1}\right)(1 + G_2 r) \qquad (3.3)$$

where G_1 is the input conductance at signal frequency, G_2 the conductance at the intermediate frequency, and B the barrier susceptance.

For practical purposes the ratio $G_1/G_2 = K$ may be taken as a constant. If it is also assumed that $r \ll 1/G_2$, it is found that, for minimum loss,

$$G_{2\text{opt}} = \frac{B}{K(1 + K)}$$

Under the above conditions, the microwave conversion loss becomes

$$L_0' = L_0 \left[1 + 2\left(\frac{1 + K}{K}\right)^{1/2} \omega Cr + \frac{1 + 2K}{K(1 + K)}(\omega Cr)^2\right] \quad (3.4)$$

It may be shown, for an ideal non-linear resistance having an exponential I/V characteristic, that $K \approx 1$ throughout. With these assumptions, it is found that, for a signal frequency of 9·375 GHz, L_0 lies in the range 3·2 − 3·5 dB for all values of C which we consider appropriate. For a frequency of 9·375 GHz, eqn. (3.4) becomes

$$L_0' = L_0[1 + 0·166Cr + 0·0883(Cr)^2] \quad (3.5)$$

with r in ohms and C in picofarads.

The results have been computed for the same combinations of layer thickness and doping level as were used for detectors, and are plotted in Fig. 3.5, which again shows the overall advantage of gallium arsenide.

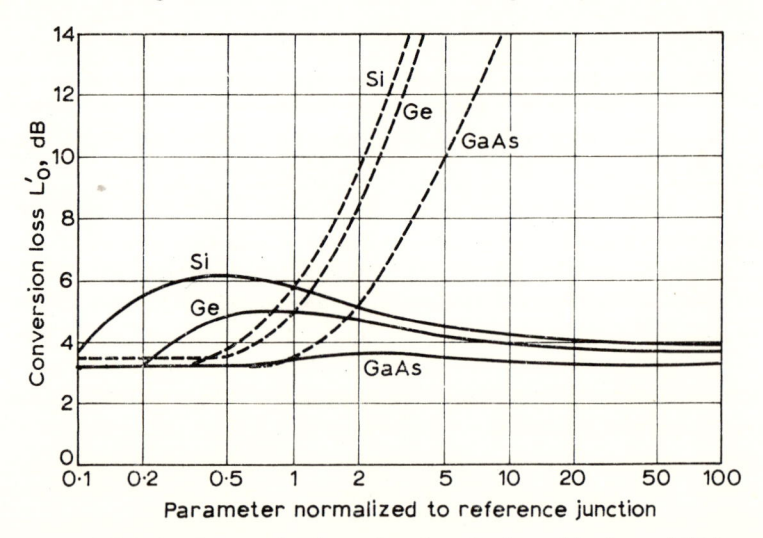

FIG 3.5 Calculated conversion losses for selected values of doping and thickness
Reference junction, 1 Ω-cm; epitaxial layer, 1 μm
——— doping level - - - - film thickness

Schottky-barrier Diodes

It is also clear that with the other materials, particularly silicon, the chosen standards of 1 μm thickness and 1 Ω-cm resistivity are not ideal. Significant improvements are indicated for thinner layers and more heavily doped material.

So far in this section we have considered only the variations in parameters relating to a particular model diode which is assumed to possess an epitaxial layer of uniform resistivity having an infinitely steep transition with the substrate. Such a layer cannot be realized in practice for the reasons previously given. We therefore consider the effect of graded transitions as predicted in Fig. 3.2, examining the effect of these profiles in relation to a model diode assumed to possess an epitaxial layer of 1 μm thickness and a junction of 25 μm diameter.

The depletion layer thickness and hence capacitance per unit area were calculated from the diffusion profiles, using the most appropriate solutions of Poisson's equation indicated in Chapter 2, with an assumed value of barrier height corresponding to tungsten. For each profile, the resistance associated with the undepleted part of the layer was found by translating the impurity profile into one of resistivity and then integrating between the depletion layer boundary and the Schottky-type contact surface. Allowance for the substrate resistance, as before, gave the total series resistance associated with the junction, thus enabling the Cr product and other device parameters to be estimated.

The results predicted from the above exercise are shown in Fig. 3.6. It is noted that, even with impurity profiles far from the ideal abrupt transition, it should be possible nevertheless to achieve tangential sensitivities greater than 61 dBm and broadband conversion losses down to 3·5 dB at 9·375 GHz. Such performances are indicated for an undoped gas stream and diffusion times of around 100 min.

In the above computations a fixed voltage bias was assumed, appropriate to optimum detector performance, but with mixers, of course, the position of the depletion layer boundary varies throughout the local oscillator driving cycle, implying a continuously varying capacitance and series resistance. However, a similar problem exists with point-contact devices, where calculations based on fixed values for these parameters nevertheless give a good guide to the performance to be expected in practice. A recent contribution to this subject has been made by Barber, who suggests the pulse duty ratio as a more fundamental parameter for defining mixer operation than the magnitude of the diode voltage [9].

Since Schottky-barrier devices offer the possibility of making metallic contacts of more robust construction than point-contact structures, an important item of design information is to know how serious is the effect of overhang capacitance on microwave performance, the "overhang" being that capacitance arising from any excess metal lying over the insulating silica film which surrounds the junction area, as described in the next section. The effect may be gauged by representing the barrier as a series

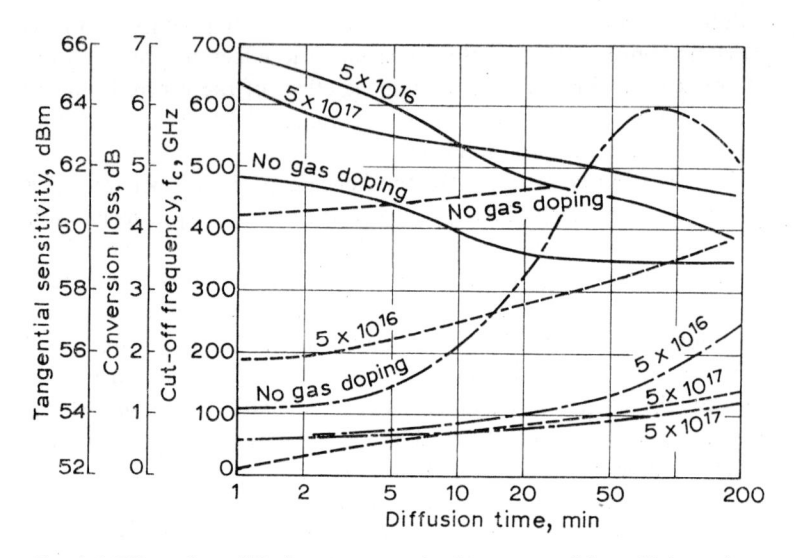

FIG 3.6 Effect of out-diffusion on conversion loss, tangential sensitivity and cut-off frequency

——— conversion loss   - - - - tangential sensitivity   —·—· cut-off frequency

combination of resistance R_1 and capacitive reactance X_1, and the overhang by a capacitive reactance X_2 in series with a resistance R_2 (attributable to the full epitaxial layer thickness plus the substrate resistance). These two branches are assumed to be in parallel, and the loss in sensitivity or increase in conversion loss will be determined by the fraction of the total conductance ascribed to the branch representing the barrier. Thus the increase in conversion loss of a mixer due to this effect is given by

$$\Delta L = 10 \log_{10} \left(1 + \frac{R_2}{R_1} \frac{R_1^2 + X_1^2}{R_2^2 + X_2^2}\right) \tag{3.6}$$

3.4 MANUFACTURE

Diodes using silicon and gallium arsenide are prepared in an essentially similar manner, and the principal techniques involve depositing metal upon the semiconductor by evaporation, by sputtering or by electrochemical means. Another method which has also been investigated is the reactive deposition of tungsten by the action of its hexafluoride on the semiconductor. In each case the following general procedures are typical:

1. A lightly doped n-type semiconductor is deposited epitaxially on to a heavily doped n-type substrate, in slice form.
2. Silica is deposited over the epitaxial layer, sometimes in two distinct stages.
3. An ohmic contact is formed to the back of the slice.
4. Oxide is removed by a photo-lithographic process from selected areas.
5. The semiconductor surface thus exposed is cleaned.
6. Metal to form the barrier is deposited, and subsequent metallic layers are built up as required for contact purposes.
7. Excess metal is removed from the oxide.
8. D.C. tests are made.
9. The prepared material is diced and encapsulated.

Oxide deposition may be effected thermally, at a temperature of around 1 200°C for silicon, or by the pyrolysis of tetraethoxysilane at a much lower temperature, say 800°C. A method of cleaning exposed areas prior to evaporation has been reported by Turner to be effective for silicon diodes [10, 11]. This involves the anodic growth of a thin layer of oxide on the surface followed by removal of this oxide immediately prior to depositing the metal. Because this technique is of considerable importance, we give details below of a particular process which has been found effective.

The silicon slice is waxed to a hollow glass holder and contact made to the ohmic back surface, prepared by electroless nickel deposition and subsequent gold plating. The slice is then immersed in a solution of boric acid and sodium tetraborate, with positive bias applied relative to a platinum cathode. This causes oxide to grow on the exposed area of silicon. On removal from the holder, the slice is given a short etch in hydrofluoric acid, which clears the anodic oxide from the holes, leaving the pryolytically deposited masking oxide substantially undamaged.

Microwave Semiconductor Devices

After the above preparation, the slice is washed and in one method of manufacture immediately transferred to an evaporator, where the barrier metal is deposited on the heated slice. With a number of metals, it is possible to wipe off the excess deposit leaving only the diode areas covered, which avoids the necessity of further photo-lithography and eliminates any overhang capacitance. An alternative method of depositing the metal

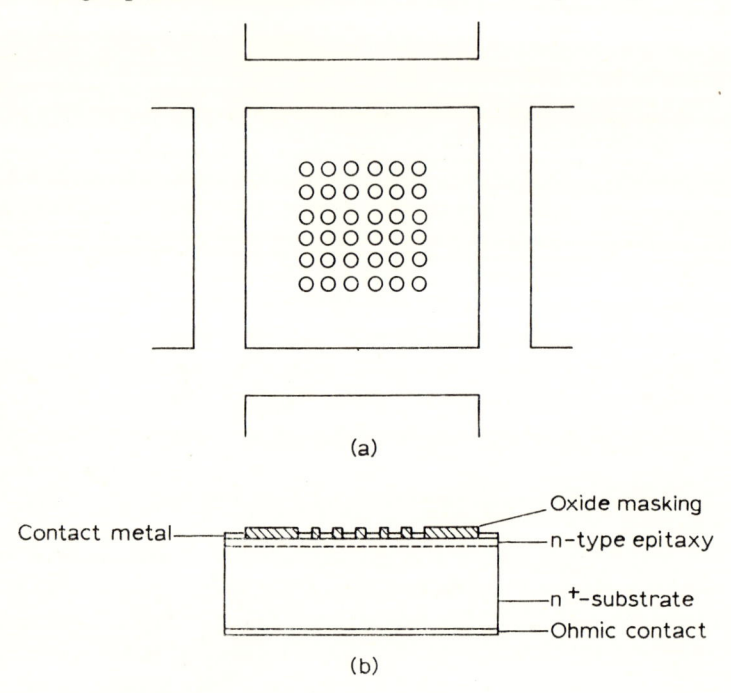

(a)

(b)

Fig 3.7 Honeycomb structure for planar junctions
    (a) Mask for random-selection contacting: 36 dots, 25 μm diameter, in each 0·63 mm square
    (b) Sectional view of diode chip

is by electroplating. Excellent results have been reported for millimetre-frequency Schottky-barrier mixers prepared in this way, using gold on gallium arsenide [1].

An attractive method of making contact to very small planar junctions is to use a "honeycomb" structure, as illustrated in Fig. 3.7. For use at wavelengths of 3 cm, diode areas of about 20 μm diameter have been found satisfactory, with pyrolytic silica layers about 1 μm thick, the pattern shown being repeated on a grid with 0·63 mm spacing. Electrical contact

Schottky-barrier Diodes

is made by allowing an electro-pointed catswhisker to skid over the silica layer until it drops into one of the holes in which a diode has been formed. It is found possible, with silicon, to substitute dice prepared as described above, instead of material normally used for point-contact diodes, and thereby to introduce Schottky-barrier structures, without significantly altering conventional point-contact assembly lines or changing the diode constructional details.

In addition to the above advantage in relation to the direct replacement of point-contact devices, there are new possibilities opened up by choice in diode area presented by the Schottky form of construction. In particular, low values of r.f. impedance, appropriate to strip-line, are readily attainable. Also, the control of r.f. and i.f. impedance through diode area is easily effected. This is not possible with point-contact devices, since their electrical parameters cannot be scaled by adjustment of the size of contact area.

Burnout is another important consideration, and the ability to withstand it is improved as the area increases. It is known that higher breakdown voltages are beneficial, too, but much further investigation of this phenomenon is required. Unfortunately, manufacturers' data can be very misleading in relation to burnout, and although values of about a microjoule

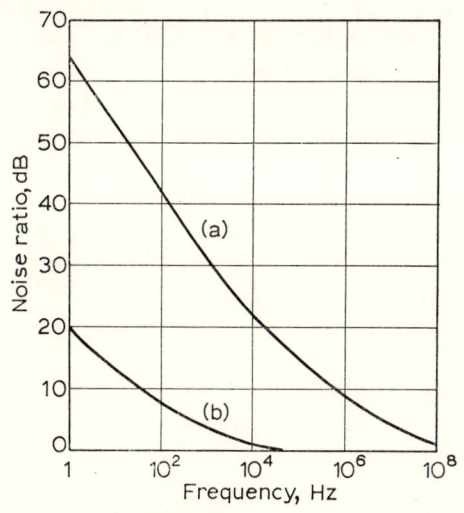

Fig 3.8 Noise comparison of typical silicon diodes at 0·5 mA bias
(a) Point-contact
(b) Schottky-barrier

107

per pulse may be quoted, it is certain that failure behind conventional gas T-R cells will occur at leakages which are only a fraction of these figures.

3.5 MEASUREMENTS

The general shapes of diode characteristics are similar to those of point-contact devices and the details depend on the material, manufacturer and diode type. One feature of Schottky-barrier diodes which is outstanding, however, is the large improvement in noise performance at low frequencies, and a comparison of such diodes with point-contact devices is illustrated in Fig. 3.8. This low-frequency noise improvement is a feature shared with the backward diode, discussed in Chapter 5.

REFERENCES

1 YOUNG, D. T., and IRVIN, J. C., "Millimeter frequency conversion using Au n-type GaAs Schottky-barrier epitaxial diodes with a novel contacting technique", *Proc. IEEE*, **53**, p. 2130 (1965).

2 CROWELL, C. R., and SZE, S. M., "Current transport in metal-semiconductor barriers", *Solid State Electronics*, **9**, p. 1035 (1966).

3 CROWELL, C. R., SHORE, H. B., and LABATE, "Surface state and interface effects in Schottky-barriers at N-type silicon surfaces", *J. Appl. P.*, **36**, p. 3843 (1965).

4 SHURMER, H. V., British Patent Applicn. 36859, 1966.

5 KAHNG, D., and D'ASARO, L. A., "Gold-epitaxial silicon high frequency diodes", *Bell Syst. Tech. J.*, **43**, p. 225 (1964).

6 CROWELL, C. R., SARACE, J. C., and SZE, S. M., "Tungsten semiconductor Schottky-barrier diodes", *Trans. Metallurical Soc. of AIME.*, **233**, p. 478 (1965).

7 BARON, C., "Theory of the microwave crystal mixer", *Proc. Instn Elect. Engrs*, **105B**, Suppl. 11, p. 662 (1958).

8 TORREY, H. C., and WHITMER, C. A., *Crystal Rectifiers* (McGraw-Hill, 1948).

9 BARBER, M. R., "Noise figure and conversion loss of the Schottky-barrier mixer diode", *Trans. IEEE*, **MTT-15**, p. 629 (1967).

10 TURNER, M. J., Ph.D.Thesis, University of Manchester (1966).

11 TURNER, M. J., and RHODERICK, E. H., "Metal-silicon Schottky barriers", *Solid State Electronics*, **11**, p. 291 (1968).

4

Tunnel Diodes

4.1 INTRODUCTION

Tunnel diodes exhibit incremental negative resistance over a small voltage range in the forward-biased direction. Their distinctive characteristic was first described in a letter from Leo Esaki appearing in the *Physical Review* for January, 1958, in which the inventor discussed the anomalous feature he had observed in the I/V characteristics of very narrow germanium p–n junctions having heavy acceptor and donor concentrations on the respective sides of the junctions, both of the order of $10^{19}/\text{cm}^3$ [1].

Other semiconductors were soon investigated in relation to the phenomenon which Esaki had observed, and of these gallium arsenide became established alongside germanium as having general utility at room temperature. Typical static characteristics for these two materials are shown in Fig. 4.1. The voltage V_p for the peak current I_p is about 60 mV with germanium and 100 mV with gallium arsenide, corresponding values of voltage V_v for the valley current I_v being 250–450 mV and 450–650 mV respectively. Most of the early work on tunnel diodes related to low-frequency applications, but subsequently they were investigated up to frequencies in the millimetre band, with resulting microwave use as oscillators, amplifiers, mixers possessing gain, and self-oscillating mixers [2].

The simple equivalent circuit for a tunnel diode is shown in Fig. 4.2. The dynamic negative resistance is believed to be independent of frequency up to well beyond the microwave range and is also substantially independent of temperature [3]. However, the capacitance shunting the negative resistance tends to be large, typically $2\,\mu\text{F}/\text{cm}^2$ of junction area. The series resistance, due to connecting leads and the semiconductor itself, is again an important factor tending to limit the high-frequency performance. Also, since tunnel diodes are essentially low-impedance devices, it is

Microwave Semiconductor Devices

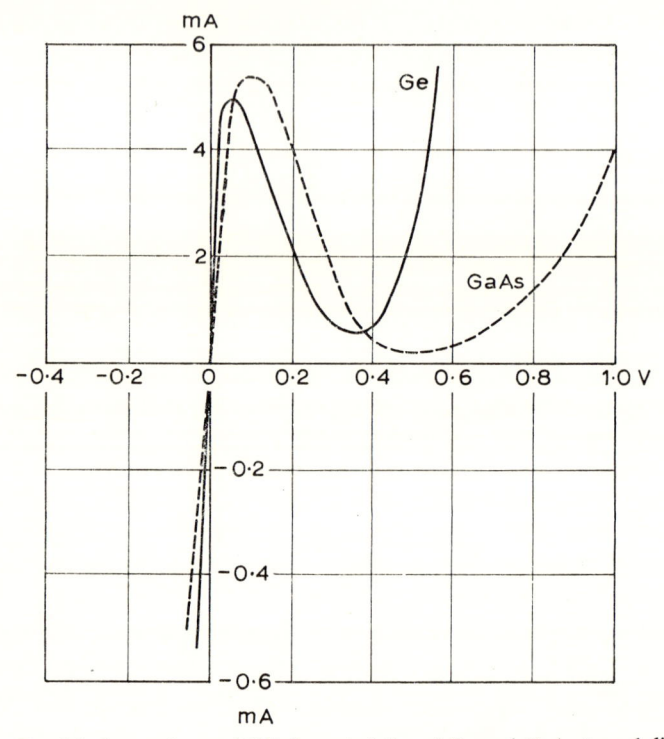

FIG 4.1 Comparison of I/V characteristics of Ge and GaAs tunnel diodes

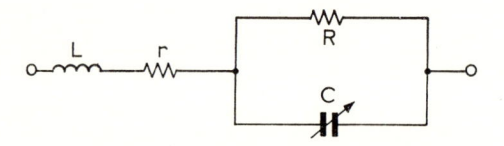

FIG 4.2 Equivalent circuit of a tunnel diode

R = negative resistance L = series inductance
C = junction capacitance r = series resistance

important in microwave applications to employ mounting arrangements which have very low series inductance.

With tunnel diodes used as microwave oscillators the power output is limited to well under 1 W, mainly on account of the relatively small voltage range over which the negative resistance applies. As amplifiers, there is a limitation arising from the two-port nature of the device, which may be overcome by using a circulator—but involving thereby increased cost and complexity. There is, however, a marked advantage with tunnel-diode mixers for certain applications owing to their modest requirements on local-oscillator power, satisfactory operation being possible down to a drive level of 100 μW. There are also applications in which the relative insensitiveness of tunnel diodes to radiation damage is important, a feature arising from the fact that minority carriers play no significant part in the operation of the device.

Among other possible tunnel diode materials, indium arsenide must be mentioned as having a high tunnelling probability, due to the low effective mass of both holes and electrons in that material. However, indium-arsenide tunnel diodes have to be cooled, preferably to 77 K [4], whereas practical upper temperature limits quoted for germanium and silicon diodes are 200°C and 350°C, respectively [5], gallium arsenide having been operated up to at least 300°C [6].

4.2 THEORY

If electron tunnelling across a *p–n* junction is to be significant, the barrier layer must be extremely thin (of the order of $10^{-2}\,\mu$m) since the tunnelling probability rises steeply with decreasing barrier thickness, as shown in Fig. 4.3. It is also necessary that there shall be electrons on one side

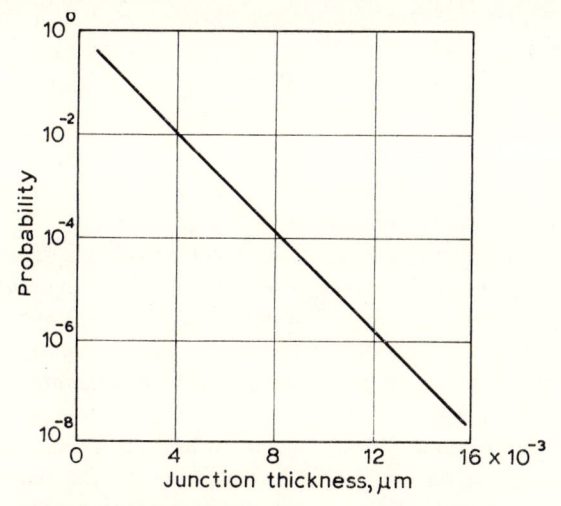

FIG 4.3 Probability of tunnelling *v.* junction thickness
(*From Ref.* 18)

Microwave Semiconductor Devices

of the junction facing empty states on the other side into which they can tunnel at the same energy level. Bearing this in mind, a simple qualitative explanation of the shape of the tunnel diode characteristic may be given with reference to Fig. 4.4.

Since the semiconductor material is so heavily doped, the Fermi level lies within a continuum of energy states either in or near the normal permitted bands. Figure 4.4(a) shows the situation in the absence of an applied voltage: at temperatures above absolute zero, electrons which are thermally excited above the Fermi level tunnel freely through the barrier at equal

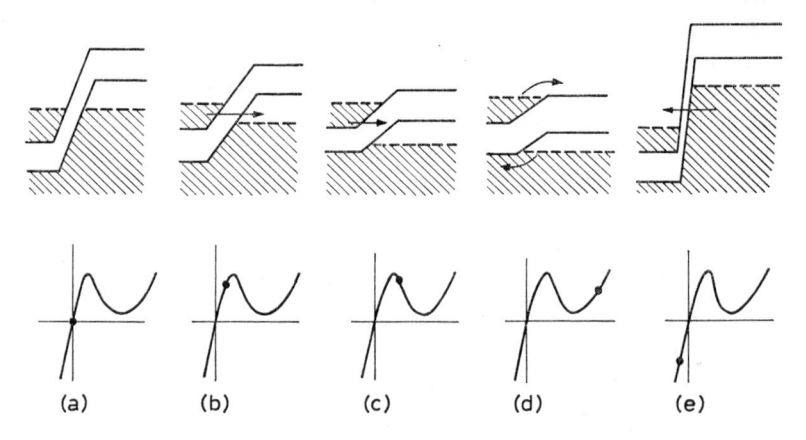

FIG 4.4 Tunnel diode characteristics: simple band picture

rates in each direction, and although the net current is zero, the individual tunnel current densities may be as high as 10^3 A/cm^2 at room temperature.

On applying a small forward bias, electrons which are at the bottom of the conduction band on the n-side of the junction become raised to energy levels lying opposite unoccupied states on the p-side, as indicated in Fig. 4.4(b). A forward current therefore flows which initially increases with bias. Eventually, however, more and more electrons become raised to levels which lie opposite the forbidden band on the p-side, into which no tunnelling is possible. When the bias reaches a certain value, enough of the electrons are in this state for the current to begin to decrease with increasing bias, which corresponds to the beginning of the negative-resistance region, as illustrated in Fig. 4.4(c). As the bias is further increased, the current remains small until minority carrier injection predominates

Tunnel Diodes

and the normal exponentially rising forward characteristic is observed, typical of a conventional *p–n* junction, as indicated at (*d*).

With reverse bias, an increasing number of electrons on the *p*-side find themselves opposite empty states in the conduction band on the *n*-side and the number of electrons which are able to tunnel increases without limit, corresponding to the situation illustrated in Fig. 4.4(*e*).

The equation for tunnelling current is written as the difference between the current from the conduction band to the valence band and that from the valence band to the conduction band:

$$I = I_{cv} - I_{vc} = KP \int_{W_c}^{W_v} \rho_c(W)\rho_v(W)[f_c(W) - f_v(W)]dW \tag{4.1}$$

where
$K = $ A constant of proportionality
$P = $ Tunnelling probability
$W_c, W_v = $ Energies of charge carries in conduction and valence bands
$\rho_c, \rho_v = $ Density of states in conduction and valence bands
$f_c(W), f_v(W) = $ Fermi–Dirac distribution functions

To a first approximation the distribution of states in both the conduction and valence bands may be assumed parabolic, in which case it may be shown [5] that the integral in eqn. (4.1) is approximated by

$$\frac{eV}{4kT}(S - eV) \tag{4.2}$$

Here S is the total penetration of the Fermi level into the bands and is the difference in electron energies between the bottom of the conduction band for the *n*-type material and the top of the valence band for the *p*-type. If it is assumed that at the junction there is a simple triangular barrier of height $\Delta W/2$, the tunnelling probability [7] is

$$P = \exp - \left[\frac{2(2\pi)^{1/2}}{3he}\bar{m}_r^{1/2}\varepsilon^{1/2}\bar{n}^{-1/2}W^{1/2}(\Delta W - eV)^{1/2}\right] \tag{4.3}$$

where $h = $ Planck's constant

$\bar{m}_r = \dfrac{\bar{m}_c \bar{m}_v}{\bar{m}_c + \bar{m}_v} = $ Reduced effective mass of charge carriers

$\bar{n} = \dfrac{n_D n_A}{n_D + n_A} = $ Reduced effective density of charge

Equation (4.3) shows that for a high tunnelling probability the effective mass of the charge carriers and the forbidden energy gap should be small whilst both the p- and n-regions should be heavily doped.

4.3 DESIGN CONSIDERATIONS

The series resistance of a tunnel diode, as with point-contact devices, is mainly due to spreading resistance at the junction together with some contribution from the contacts. The negative resistance, to a first approximation, is inversely proportional to the product of junction area and tunnelling probability. The junction capacitance is almost entirely due to space-charge depletion, there being very little of the storage capacitance associated with the diffusion of minority carriers.

It follows from the above data that the product of negative resistance $|R|$ and junction capacitance C, which mainly limits the upper frequency of operation, is a decreasing function of doping level, and it is stated by Sommers [2] that for germanium diodes the $|R|C$ product decreases from 5×10^{-9}s to 5×10^{-11}s on increasing the doping of the semiconductor from 2·4 to 4·8 $\times 10^{19}$/cm³.

Referring to Fig. 4.2, the small-signal impedance Z across the terminals of a tunnel diode, biased into its negative resistance region, is given by

$$Z = \left(r - \frac{|R|}{1 + \omega^2 C^2 R^2}\right) + j\left(\omega L - \frac{\omega C R^2}{1 + \omega^2 C^2 R^2}\right) \quad (4.4)$$

In order that the real part of this impedance shall be negative the frequency must be less than a cut-off value f_c given by

$$f_c = \frac{1}{2\pi |R| C} \left(\frac{|R|}{r} - 1\right)^{1/2} \quad (4.5)$$

The imaginary part becomes zero at a frequency f_r, known as the self-resonant frequency of the diode, given by

$$f_r = \frac{1}{2\pi} \left(\frac{1}{LC} - \frac{1}{C^2 R^2}\right)^{1/2} \quad (4.6)$$

4.3.1 Stability

Consider a steady voltage V_b to be applied across the equivalent circuit of Fig. 4.2, except that we shall replace R in the equivalent circuit by the static junction resistance R_s and take r to include the load resistance and

Tunnel Diodes

circuit losses. The diode capacitance will be assumed to be represented approximately by

$$C \approx K(V_0 - V)^{1/2}$$

where $V_0 \approx 0.6\,\text{V}$ for silicon and $1.1\,\text{V}$ for gallium arsenide. Application of Kirchhoff's laws then leads to the simultaneous differential equations

$$V + KR_s(V_0 - V)^{1/2}\frac{dv}{dt} - IR_s \approx 0 \qquad (4.7)$$

$$Ir + L\frac{di}{dt} + V = V_b \qquad (4.8)$$

where V represents the voltage appearing across the parallel combination of R_s and C, and I is the current through L and r.

To proceed with the analysis it is assumed that R_s and C are substantially constant and equal to their initial values R_{si} and C_i for small voltage excursions. The initial conditions relating to eqns. (4.7) and (4.8) are chosen to make the dynamic resistance R negative at a time $t = 0$. The Laplace transformation of these equations then gives the initial response of the circuit to a small excitation by the characteristic equation:

$$s^2 L C_i + s\left(rC_i - \frac{L}{|R_{si}|}\right) + 1 - \frac{r}{|R_{si}|} = 0 \qquad (4.9)$$

Solving this equation for s, the complex angular frequency [8],

$$s_{1,2} = -\frac{1}{2}\left(\frac{r}{L} - \frac{1}{C_i|R_{si}|}\right) \pm \left[\frac{1}{4}\left(\frac{r}{L} - \frac{1}{C_i|R_{si}|}\right)^2 - \frac{1 - r/|R_{si}|}{LC_i}\right]^{1/2}$$

$$= \sigma_0 \pm j\omega_0 \qquad (4.10)$$

The circuit will be unstable if $\sigma_0 > 0$ and stable if $\sigma_0 < 0$. We here quote results given by Nelson [9] for three cases of practical interest, in each of which $r < |R|$ but with additional conditions as follows:

(1) $\qquad L < r_s|R|C \quad (r_s = \text{diode series resistance}) \qquad (4.11)$

The diode is short-circuit stable and suitable for use as an amplifier or as an oscillator if sufficient reactance is added to the circuit.

(2) $\qquad r_s|R|C < L < |R|^2 C \qquad (4.12)$

Microwave Semiconductor Devices

The diode is conditionally stable and may be used as an amplifier if the load resistance is sufficiently large that

$$L < r|R|C \tag{4.13}$$

and may be used as an oscillator if this condition is not satisfied.

(3) $\quad L > |R|^2 C$

The diode is unstable and the circuit cannot be made stable for amplifier use by adding resistance, but may still be used as an oscillator.*

4.3.2 Amplifiers

Gain
Consider the simplest form of amplifier shown in Fig. 4.5, in which a tunnel diode in parallel with a transmission line is tuned by an inductance

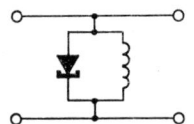

FIG 4.5 Simple tunnel diode amplifier

L, and assume that one or other of the above conditions for amplification is satisfied. The available power gain at resonance is defined as

$$G_p = \frac{\text{Power supplied to load}}{\text{Maximum available power from generator}}$$

and is readily shown to be given by

$$G_p = \frac{4 g_L g_G}{g_L + g_G + g_D} \tag{4.14}$$

where g_L, g_G and g_D are respectively the conductances of the load, generator and diode appropriate to the resonant frequency ω_0 as indicated in Fig. 4.6.

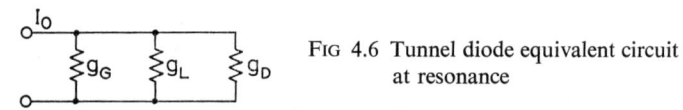

FIG 4.6 Tunnel diode equivalent circuit at resonance

It is to be noted that the total positive conductance must exceed the diode negative conductance if oscillations are to be avoided.

* The above results are reproduced from *Microwave Solid-state Engineering* by L. S. Nergaard and M. Glicksman (Litton Educational Publishing, Inc., 1964), with permission.

Tunnel Diodes

At any frequency ω other than resonance the power gain is given by

$$G_p(\omega) = \frac{4g_G g_L}{(g_L + g_G + g_D)^2 + \omega^2 C^2(1 - \omega_0/\omega)^2} \tag{4.15}$$

Now, the amplifier bandwidth B is defined as $1/2\pi$ times the difference between the two angular frequencies ω_1 and ω_2 at which the gain drops to half of its value at resonance. Thus

$$B = \frac{\omega_2 - \omega_1}{2\pi} = \frac{g_G + g_L + g_D}{2\pi C} \tag{4.16}$$

where C is the effective diode capacitance in parallel with the transmission line.

It will be observed that, as the total positive conductance approaches the numerical value of the diode negative conductance, the circuit Q-factor increases with resulting increase in gain and decrease in bandwidth. Since a compromise has to be made between gain at resonance and bandwidth, it is convenient to take as a figure of merit for the amplifier the gain-bandwidth product. If we here take the gain as $\sqrt{G_p}$ we have

$$\sqrt{G_p} B = \frac{\sqrt{(g_G g_L)}}{\pi C} \tag{4.17}$$

The maximum value of this product is given by

$$(\sqrt{G_p} B)_{max} = \frac{g_D}{2\pi C} \tag{4.18}$$

which indicates that the figure of merit of the amplifier can be given in terms of the tunnel diode parameters alone. Clearly the ratio g_D/C should be as high as possible, and this is also the condition necessary for high cut-off frequency, as will be seen on inspection of eqn. (4.5). The limitation indicated by eqn. (4.18) does not apply to circuits of greater complexity, as shown by Seidel and Herrmann [10].

Noise Figure

The amplifier noise figure is derived with reference to Fig. 4.6, which indicates the three effective noise sources, i.e. the diode shot noise, generator thermal noise and load thermal noise. The corresponding mean square currents are respectively given by the following equations:

$$\overline{I_D{}^2} = 2eI\delta f \tag{4.19}$$

$$\overline{I_G{}^2} = 4kT_G g_G\, \delta f \tag{4.20}$$

$$\overline{I_L{}^2} = 4kT_L g_L\, \delta f \tag{4.21}$$

where T_G and T_L represent respectively the absolute temperatures of the generator and load conductances. It should be noted that in eqn. (4.19) I is the sum of forward and reverse currents flowing in the diode.

Taking the noise figure F as the ratio of the sum of noise powers delivered to the load conductance by the three noise generators to the contribution of the source generator alone, we find that

$$F = 1 + \frac{T_L}{T_G}\left(\frac{g_L}{g_G}\right) + \frac{eI}{kT_G g_G} \tag{4.22}$$

It is clear from this equation that minimum noise figure will be achieved when the load conductance approaches zero and the generator conductance is high. Clearly a small load conductance is in conflict with the requirement for high gain-bandwidth product expressed by eqn. (4.17). However, for a high-gain amplifier it is implicit that $g_G \approx g_L$, in which case the minimum noise figure attainable is given by

$$F_{\min} = 1 + \frac{eI}{2kT_G g_D} = 1 + 20\,\frac{I}{g_D} \tag{4.23}$$

At the optimum gain-bandwidth setting, with $g_L = g_G = g_D/2$, it is to be noted from eqn. (4.22) that the amplifier noise figure, with generator and load at the same temperature, is given by

$$F = 2\left(1 + \frac{eI}{2kT_G g_D}\right) = 2F_{\min} \tag{4.24}$$

In practice, a compromise has to be made between gain-bandwidth product and noise figure. A detailed account of the problems encountered is given in Ref. 11.

4.3.3 Oscillators

The conditions for oscillation are obtained from the stability criteria. Equations. (4.11)–(4.13) indicate that these conditions are not in general

Tunnel Diodes

difficult to achieve. For steady-state oscillations the sum of the effective impedances or admittances at any point in the circuit must be zero, and it may be shown [12] from eqn. (4.10) that, provided only moderate variations occur in the capacitance and resistance of the junction, we have

$$L \approx r|\bar{R}|\bar{C} \tag{4.25}$$

$$\omega_s \approx \left(\frac{1 - r/|\bar{R}|}{L\bar{C}}\right)^{1/2} \tag{4.26}$$

where L and r now represent the total circuit series inductance and resistance, ω_s is the steady-state angular frequency, and $\bar{R}$ and $\bar{C}$ are the average effective values of the diode resistance and capacitance.

In terms of the r.m.s. values of voltage and current, V_r and I_r, the oscillatory power delivered by a negative resistance is, of course, given by

$$P = V_r I_r \tag{4.27}$$

Assuming, to a first approximation, that the tunnel diode is a linear negative resistance in the range V_p to V_v, the power delivered to the diode (assuming negligible series resistance), when the r.f. voltage swing is confined to this range, is given by

$$P \approx \tfrac{1}{8}(V_v - V_p)(I_p - I_v) \tag{4.28}$$

It is found in practice that a better approximation is obtained by changing the constant in eqn. (4.28) from $\tfrac{1}{8}$ to $\tfrac{3}{16}$.

Complete solutions of eqns. (4.7) and (4.8) (e.g. power output, harmonic content) can be obtained by numerical calculus or by a combination of graphical and analytical procedures [13].

4.4 MANUFACTURE

The abrupt junctions required for tunnel diodes are most readily formed by alloying. Probably the simplest means of making a germanium tunnel diode is to alloy a pellet of indium into heavily doped n-type material. Some of the earliest commercial devices were made in this way, with a bridge-type contact to the alloyed dot, the contact area being restricted by etching around the dot to form a mesa structure. Such devices potted in epoxy are suitable for u.h.f. switching and converter applications.

Microwave Semiconductor Devices

For microwave applications, more sophisticated structures have been developed. In one arrangement a sandwich form of construction is employed, in which slices of germanium, one n-type and one p-type, separated by a disc of glass, are bonded together. The sandwich is then cut into square wafers of about 0·5 mm sides and the glass etched back slightly. An arsenic-doped pellet of lead or lead-indium is alloyed to both germanium wafers and subsequently etched before the unit is assembled either with strip leads or in a "pill" package. An alternative arrangement uses an "inverse sandwich" with a single germanium wafer bonded between two glass plates. After cutting into 0·5 mm strips dumbbell patterns of lead–indium–arsenic are evaporated and alloyed, the width of the dumbbell being subsequently reduced by etching to a few microns.

Microwave tunnel diodes of a few milliamperes peak current are available with resistive cut-off frequencies up to about 100 GHz. Series resistance is usually in the range 3–10 Ω; junction capacitance, 0·1–2·0 pF; and negative resistance, around 50 Ω.

4.5 MICROWAVE MOUNTING

The inductance of the package presents a problem at microwave frequencies, since tunnel diodes are essentially low-impedance devices. A possible solution is to avoid a package where possible, and one such arrangement, described by Sommers [2], is illustrated in Fig. 4.7. In this microstrip

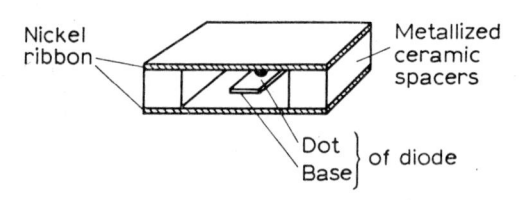

Fig 4.7 Tunnel diode mounted in microstrip
(*From Ref. 2*)

system, $\frac{1}{8} \times \frac{3}{8}$ in nickel ribbons are bonded on either side of ceramic spacers, which measure 10 thou by $\frac{1}{8} \times \frac{1}{8}$ in. In the $\frac{1}{8}$ in opening between the ceramics the diode is mounted with its base soldered to the bottom electrode and the alloyed dot to the upper one, the opening being filled with epoxy for mechanical strength. The unit is designed to be clamped directly into a section of microstrip transmission line.

Tunnel Diodes

A major problem in tunnel diode circuits is to suppress unwanted modes, and one method described by Sommers for doing this is to use a non-inductive resistor at a point so close to the diode that the intervening inductance is unimportant compared to the effective inductance of the desired mode—the point of connection being one at which the voltage swing for the wanted mode is zero. Such a configuration is illustrated schematically in Fig. 4.8(*a*). The tunnel diode is at D, its low impedance permitting it to be close to the zero voltage point at C.

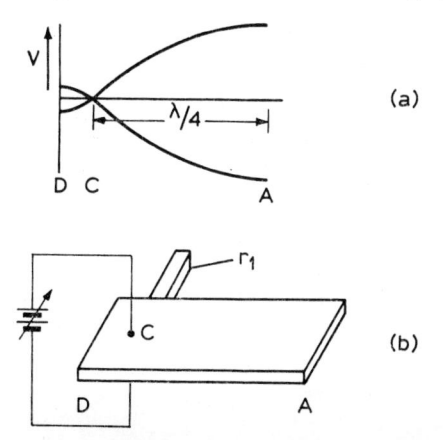

FIG 4.8 Method for suppressing unwanted modes (*From Ref.* 2)

In Fig. 4.8(*b*) is shown the microstrip arrangement of diode clamped between the two conducting foils at D, there being an open-circuit at A. The non-inductive resistor is shown connected at C in a short T-arm. The d.c. connection can also be made at C without interfering with the r.f. mode. This particular oscillator has also the advantage of giving a voltage step-up from the diode to the open end of the line, thus facilitating matching into standard impedance line.

Tunnel diodes are also used in cavity-type mounts. Figure 4.9 illustrates schematically a coaxial microwave cavity of a type used by Yariv and Dickten of Bell Telephone Laboratories and described by Hines [14], giving oscillations in the 3 cm band. Strip-line cavities have also been used with success on both amplifier and oscillator applications. Burrus and Trambarulo [15, 16] have obtained oscillations and amplification at millimetre wavelengths using point-contact tunnel diodes in a ridge-waveguide mount, Fig. 4.10.

Microwave Semiconductor Devices

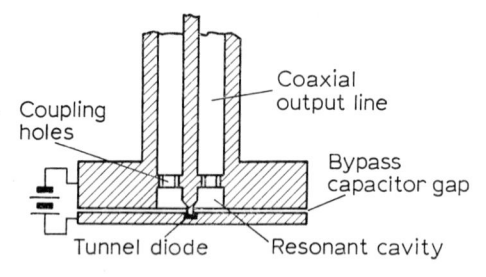

FIG 4.9 Coaxial cavity tunnel diode mount
(*From Ref.* 14)

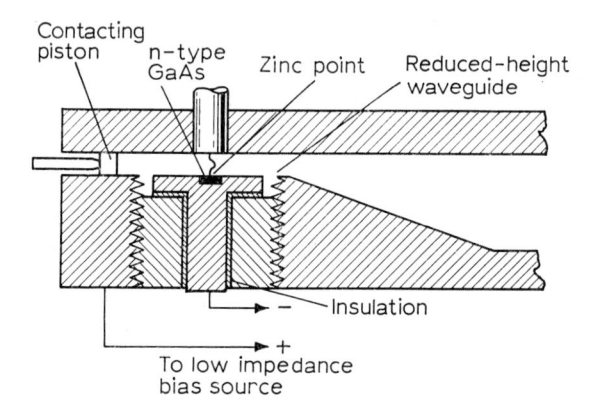

FIG 4.10 Millimetre-wave tunnel diode amplifier
(*From Ref.* 16)

4.6 MICROWAVE MEASUREMENTS

The microwave properties of tunnel diodes are determined through the interpretation of v.s.w.r. measurements. Discussion of the somewhat sophisticated methods which have been developed is outside the scope of this book but may be found in Refs. 11 and 17.

REFERENCES

1 ESAKI, L., "New phenomenon in narrow germanium *p–n* junction", *Phys. Rev.*, **109**, p. 663 (1958).

2 SOMMERS, H. S., "Tunnel diodes as high-frequency devices", *Proc. Inst. Radio Engrs*, **47**, p. 1201 (1959).

3 GARTNER, W. W., "Esaki or tunnel diodes—Part 1", *Semiconductor Products*, p. 31 (1960).

4 BATDORF, R. L., DACEY, G. C., WALLACE, R. L., and WALSH, D. J., "Esaki diode in InSb", *J. Appl. Phys.*, **31**, p. 613 (1960).

5 PRZYBYLSKI, J., and ROBERTS, G. N., "The design and construction of tunnel diodes", *J. Brit. Instn Radio Engrs*, **22**, p. 497 (1961).

6 HOLONYAK, N., and LESH, I. A., "Gallium arsenide tunnel diodes", *Proc. Inst. Radio Engrs*, **48**, p. 1405 (1960).

7 CONDON, E. U., and MORSE, P. M., "Influence of the lattice vibrations of a crystal on the production of electron-hole pairs in a strong electric field", *Rev. Mod. Phys.*, **3**, p. 43 (1931).

8 MARTIN, T. L., *Electronic Circuits*, Chap. 2 (Prentice-Hall, Englewood Cliffs, N.J., 1955).

9 NELSON, D. E., *Microwave Solid State Engineering*, Chap. 4 (Van Nostrand, Princeton, N.J., 1964).

10 SEIDEL, H., and HERRMANN, G. F., "Circuit aspects of parametric amplifiers", *Inst. Radio Engrs Wescon Conv. Rec.*, Pt. 2, p. 53 (1959).

11 KIM, C. S., and LEE, C. W., "Microwave measurement of tunnel diode parameters", *Microwaves*, **3**, No. 11, p. 18 (1964).

12 STERZER, F., and NELSON, D. E., "Tunnel diode microwave oscillators", *Proc. Inst. Radio Engrs*, **49**, p. 744 (1961).

13 STRAUSS, L., *Wave Generation and Shaping*, Chap. 15 (McGraw-Hill, 1960).

14 HINES, M. E., "High-frequency negative-resistance circuit principles for Esaki diode applications", *Bell Syst. Tech. J.*, **39**, p. 477 (1960).

15 BURRUS, C. A., "Millimeter wave Esaki diode oscillators", *Proc. Inst. Radio Engrs*, **48**, p. 2024 (1960).

16 BURRUS, C. A., and TRAMBARULO, R., "A millimeter-wave Esaki diode amplifier", *ibid.*, **49**, p. 1075 (1961).

17 FUKUI, H., "The characteristics of Esaki diodes at microwave frequencies", *J. Inst. Elect. Commun. Engrs, Japan*, **43**, p. 1351 (1960).

18 MROSEIWICZ, B., and HEASELL, E., *J. Electron. Control*, **10**, No. 6, p. 405 (1961).

5

Backward Diodes

5.1 INTRODUCTION

A backward diode is a special form of tunnel diode in which the tunnelling is incipient only and the negative-resistance region virtually disappears. The forward current is then very small and becomes equivalent to the reverse current of a conventional diode. The shape of the I/V characteristic is compared with those of point-contact and tunnel diodes in Fig. 5.1.

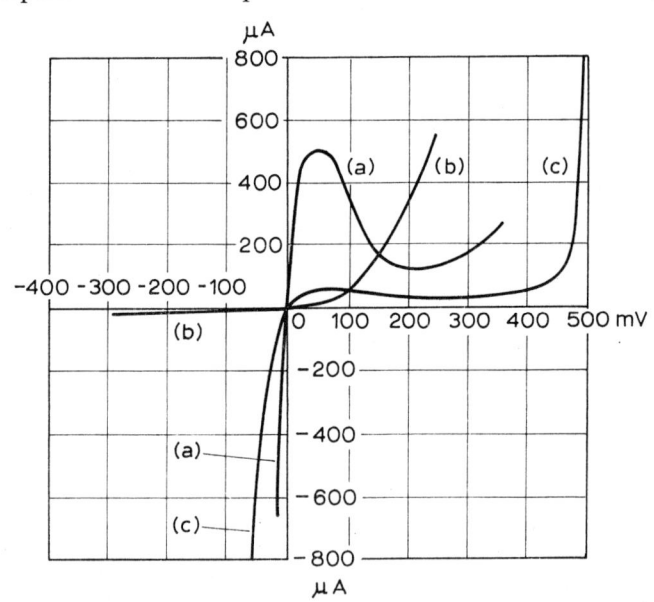

FIG 5.1 D.C. characteristics of tunnel diodes, backward diodes and point-contact diodes
(a) Tunnel diode
(b) Point-contact diode
(c) Alloyed backward diode

124

Backward Diodes

Backward diodes may be used as detectors which have high sensitivity at zero bias, or as mixers requiring only a relatively low level of local oscillator excitation. Both as detectors and mixers they may be employed in applications where flicker noise must be kept to a minimum. Low noise is an attribute of the quantum-mechanical tunnelling process and is facilitated by the low drive requirements.

For detector crystals the best type of characteristic is one which bends most sharply at the origin, and Fig. 5.1 shows that the backward diode is more suitable in this respect than a true tunnel diode. The latter can give high sensitivity in the neighbourhood of its peak and valley currents, but apart from requiring bias to achieve this, the resulting dynamic impedances are altogether too high if the junction areas are kept small enough to satisfy the requirements on capacitance. Although doping levels for backward diodes are less than for tunnel diodes, the resistivity of the semiconductor material employed is nevertheless an order of magnitude below that for conventional point-contact devices. Since low spreading resistance thereby results, little degradation occurs in current sensitivity and conversion loss from l.f. values up to at least 10 GHz.

A practical way of fabricating backward diodes with junction areas small enough for use at microwave frequencies is through a technique originally described by Eng [1], who formed backward diodes by pulse-bonding gallium-plated gold wires to degenerate n-type germanium. More recently it has proved possible to make suitable alloyed structures by planar techniques.

Alloyed structures tend to have larger capacitances than do junctions which are formed by pulse-bonding. However, they have the advantages of greater mechanical rigidity and are inherently of lower impedance, thus being more compatible with integrated circuit technology. A feature common to both types is their relative insensitiveness to temperature changes. Their principal disadvantages are in respect of limited dynamic range and modest burnout performance.

5.2 DETECTORS

The figure of merit M for detectors has already been defined in relation to signal/noise voltage ratio in eqn. (1.10), and the tangential sensitivity, for a receiver of 1 MHz bandwidth, is given in terms of M by eqn. (1.12). Since the latter is a function of video impedance and current sensitivity only, for a given receiver, it is possible to plot a theoretical curve of low-frequency tangential sensitivity versus video impedance, taking for β_0 the value derived in Section 5.2.1. In Fig. 5.2 the resulting curve is compared with experimental values quoted by Oxley and Hilsden [5] for frequencies of approximately 10 and 30 GHz.

It is to be noted that at 10 GHz the agreement between the experimental and theoretical curves is good, indicating that the degradation factor in r.f. current sensitivity β compared with β_0 is small at this frequency [see eqn. (1.8)]. At 30 GHz, however, the measured tangential sensitivity is

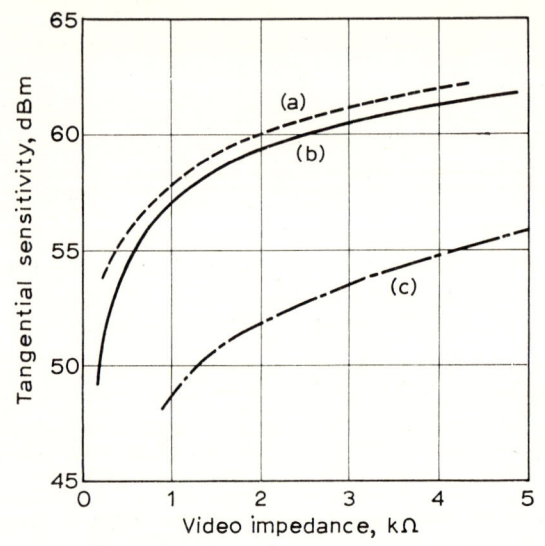

Fig 5.2 Tangential sensitivity as a function of video impedance (1 MHz video bandwidth)
(a) Theoretical
(b) Experimental, c . 10 GHz
(c) Experimental, c . 30 GHz
(From Ref. 5)

worse than the ideal theoretical value by about an order of magnitude over most of the video impedance range which is shown, the discrepancy becoming less as the video impedance increases.

In video receiver applications, wide r.f. bandwidths are normally required. This is facilitated by making the detector video impedance low, i.e. of the same order as the feeder impedance, which demands some compromise in sensitivity. With backward diodes, unlike conventional point-contact detectors, the video impedance can be made to approach the required low values even in the absence of forward bias, whilst

Backward Diodes

maintaining sensitivity values at least as good as for the more conventional detectors. A further aid to wide bandwidth is low capacitance, which implies a smaller junction area, but usually involves some sacrifice in burnout capability.

Forward bias may be applied to backward diodes which in an unbiased condition have relatively high levels of video impedance. In this way

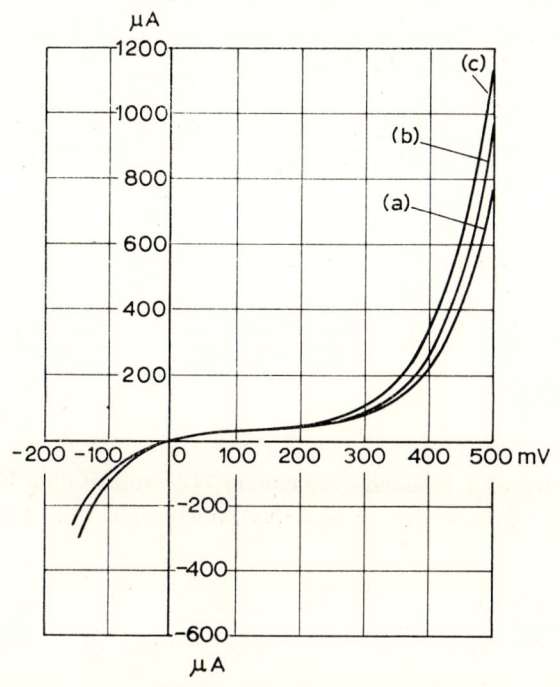

FIG 5.3 Variation with temperature of characteristics for pulse-bonded point-contact backward diodes
(a) 22°C (b) 44°C (c) 89°C

reduced video impedance may be achieved without excessive loss in sensitivity, but unbiased backward diodes of the desired low video impedance are to be preferred.

Backward diodes are relatively insensitive to variations in temperature, and Fig. 5.3 shows the changes in I/V characteristics with temperature for pulse-bonded diodes. Oxley and Hilsden [5] have quoted a reduction in tangential sensitivity of about 2dB at 150°C compared with the value at room temperature.

Microwave Semiconductor Devices

Backward diode detectors have been shown to be less susceptible to burnout than conventional detectors for video receiver applications, in which the primary requirement is the ability to retain a constant r.f. admittance after exposure to pulses of about $1\,\mu s$ duration for relatively long periods [6].

5.2.1 Theoretical Derivation of L.F. Current Sensitivity β_0

It will be understood that the basic theory underlying tunnel diodes given in Chapter 4 applies to backward diodes also. There is, however, one electrical parameter which it is of interest to examine theoretically, and this is the l.f. current sensitivity β_0. The result, derived below, is found to be in good agreement with measured values.

If the tunnelling current is expressed by eqn. (4.1) and the approximation of eqn. (4.2) applies, then assuming the validity of eqn. (4.3), the I/V characteristic of any junction in which tunnelling predominates is, over a limited range, of the form

$$I = c_1 V(S - eV)^2 \exp - [c_2(\Delta W - eV)^{1/2}] \tag{5.1}$$

where $c_1 = Ke/4kT$

$$c_2 = \frac{2(2\pi)^{1/2}}{3he} \bar{m}_r^{1/2} \epsilon_s^{1/2} \bar{n}^{-1/2} \Delta W^{1/2}$$

The zero bias value of β_0 is found from eqn. (1.7) on substituting $V = 0$ in the first and second derivatives of eqn. (5.1), obtained as follows:

First derivative

$$f'(v) = c_1(eV - S) \exp - [c_2(\Delta W - eV)^{1/2}]$$
$$[(3eV - S) + \tfrac{1}{2}c_2 eV(eV - S)(\Delta W - eV)^{-1/2}]$$
$$'(o) = c_1 S^2 \exp - [c_2 \Delta W^{1/2}] \tag{5.2}$$

Second derivative

$$f'' = eC_1 \exp - [C_2(\Delta W - eV)^{1/2}]$$
$$\Big\{2(3eV - 2S) + C_2(\Delta W - eV)^{-1/2}$$
$$(3eV - S)(eV - S) + \frac{C_2 eV}{4}(eV - S)^2$$
$$(\Delta W - eV)^{-1}[(\Delta W - eV)^{-1/2} + C_2]\Big\}$$

$$f''(o) = eSC_1 \exp - (C_2 \Delta W^{1/2})(C_2 S \Delta W^{-1/2} - 4) \tag{5.3}$$

Backward Diodes

Taking the values appropriate to $V = 0$ for the two derivatives, eqn. (1.7) leads to the result

$$\beta_0 = \exp\left(\frac{C_2 \Delta W^{-1/2}}{2} - \frac{2}{S}\right) \tag{5.4}$$

In evaluating C_2 we may use values for the various constants given by Wright [2]; thus $\bar{m}_r = 6\cdot 83 \times 10^{-29}$ g, with $\epsilon_s = 15\cdot 7$; also $\Delta W = 0\cdot 65$ eV. Spitzer *et al.* [3] quote $n_D = 2 \times 10^{19}$ carriers/cm^3 as corresponding to phosphorus-doped germanium with a resistivity of $0\cdot 001\,\Omega$-cm. Assuming that a similar value for n_A is obtained on the p-type side of the junction, then $\bar{n} = 10^{19}$ carriers/cm^3. This assumption cannot be fully justified, but errors in $\bar{n}$ and $\bar{m}$ caused by asymmetry of the junction are not likely to be too serious.

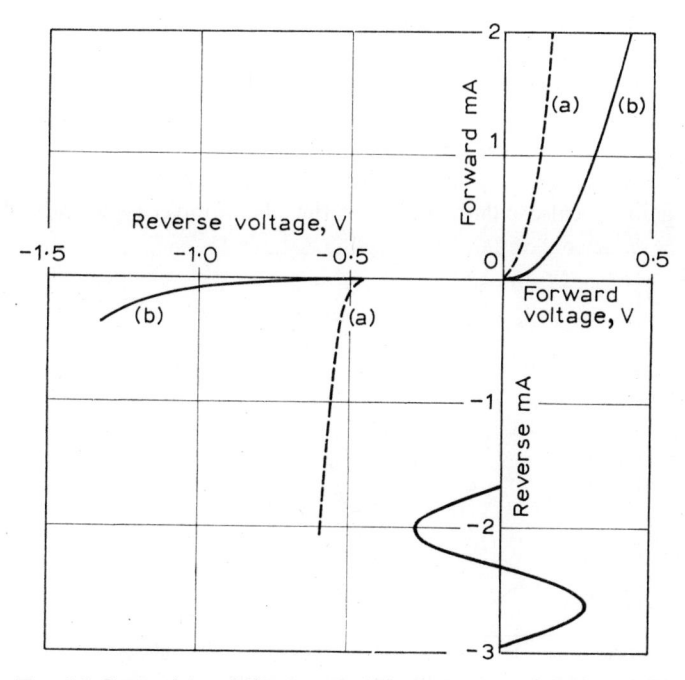

FIG 5.4 Comparison of I/V characteristics for backward diode and silicon mixer in relation to local-oscillator drive

(a) Backward diode
(b) Silicon mixer CV2154/5
(*From Ref.* 5)

The resulting value for the first term in eqn. (5.4) is $4\cdot3\,\text{V}^{-1}$. Taking Esaki's value [4] for S, which applies to concentrations on each side of the junction which are of the order of 10^{19} carriers/cm^3, the second term has the value $16\cdot1\,\mu\text{A}/\mu\text{W}$. Thus $\beta_0 = (4\cdot3 - 16\cdot1)\,\text{V}^{-1}$, i.e. $11\cdot8\,\mu\text{A}/\mu\text{W}$, ignoring the sign (which is, of course, positive when the usual convention of polarity is followed).

It is interesting to note that for gallium arsenide the product $C_2\Delta W^{-1/2}$ in eqn. (5.4) is less than for germanium but S is considerably greater, and the net result is a prediction of inferior values of β_0.

5.3 MIXERS

A typical local-oscillator voltage swing is compared in Fig. 5.4 with characteristics of backward-diode and conventional point-contact mixers. The steeply rising forward characteristic of the backward diode is indicative of good rectification efficiency, associated with low conversion loss. However, the sharp rise in reverse current at a relatively small reverse voltage also indicates a limited dynamic range compared with the point-contact mixer, since for drive levels exceeding this value the noise temperature ratio rises rapidly and the conversion loss also increases.

Figure 5.5, due to Oxley and Hilsden, shows the variation of overall noise factor against local-oscillator drive for the silicon mixer type CV2154 and

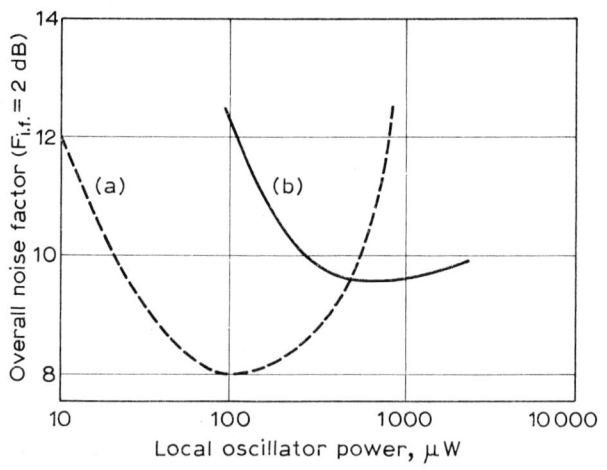

FIG 5.5 Overall noise factor $v.$ local oscillator drive (i.f., 45 MHz)
(a) Backward diode
(b) Silicon mixer CV2154/5
(From Ref. 5)

a germanium bonded backward diode. It is to be noted that the minimum value occurs at about 100 μW for the latter device and that it is superior down to a drive level as low as 30 μW. This is of advantage in permitting the use of solid-state sources of low output power for local oscillator duty.

The dynamic range, expressed as the permissible variation of local-oscillator drive power, which is available from a backward diode may be modified during manufacture. This is effected at the pulsing or alloying stage, according to the type of diode, through control of the degree of tunnelling. A further result, dependent on this, is illustrated in Fig. 5.6,

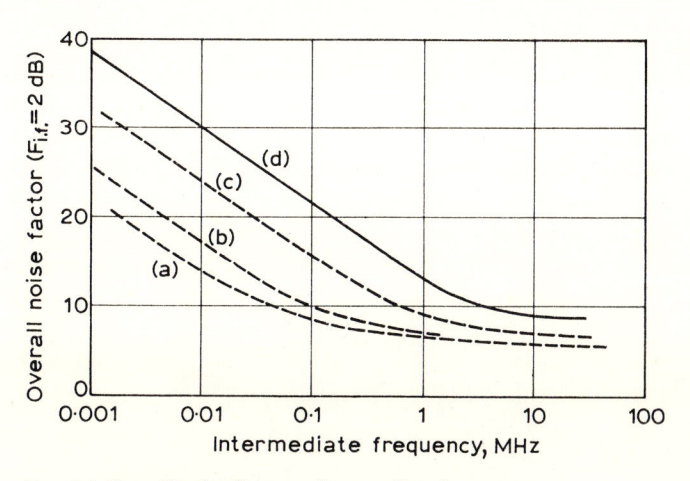

FIG 5.6 Overall noise factor $v.$ intermediate frequency
    (a) Backward diode, $I_p = 250$ μA
    (b) Backward diode, $I_p = 30$ μA
    (c) Backward diode, $I_p = 0$
    (d) Silicon mixer CV2154/5
    (From Ref. 5)

which reproduces curves, again due to Oxley and Hilsden, of overall noise factor as a function of intermediate frequency. Backward diodes exhibiting varying degrees of tunnelling are classified by the height of the incipient hump in their I/V characteristic. It is to be noted that at 1 kHz a backward diode with insufficient tunnelling to give a hump ($I_p = 0$) is superior in overall noise factor by about 6 dB over a conventional point-contact mixer type CV2154. The superiority increases with the degree of tunnelling: for a diode with $I_p = 30$ μA the improvement is some 13 dB; and for $I_p = 250$ μA it is approximately 16 dB. For diodes of the latter type the degradation at 10 kHz is only about 5 dB compared with the asymptotic value which is reached at frequencies beyond 200 kHz.

Backward-diode mixers also show a marked superiority over conventional point-contact mixers at high temperatures. Typically, the deterioration in overall noise factor between 20°C and 100°C is only about 0·5 dB compared with 1·5 dB for the CV2154. With regard to burnout, the ability of backward diodes to withstand "spikes" of energy overload is very similar to that of point-contact mixers if the junction areas are not too small.

5.4 MANUFACTURE

The pulse-bonding technique described by Eng involves the application of a pulse through a gallium-plated gold wire brought into contact with an n-type semiconductor wafer of resistivity in the range 0·001–0·004 Ω-cm. The pulsing equipment provides pulses of variable height and width. By viewing the I/V characteristic on an oscilloscope the junction is pulsed

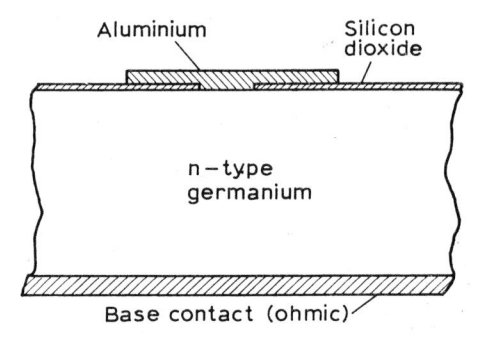

Fig 5.7 Planar form of backward diode

until the desired curve is obtained. This method has subsequently been adopted by others [6, 7].

The writer has obtained best results with germanium having lower resistivity than quoted above, around 0·0005 Ω-cm, phosphorus proving more suitable as a doping agent than arsenic. A suitable surface preparation procedure is to degrease the mounted slice in boiling acetone before etching for 2 min in a mixture of nitric and hydrofluoric acids, followed by conventional cleansing operations. Further findings were that gallium-plated gold was among the best metals investigated with regard to electrical

Backward Diodes

characteristics, and pure gold appeared superior to a 1% gallium-gold alloy. Silver appeared to be just as good and had an advantage in hardness. Alloys of 1% and 5% gallium in silver also gave encouraging results, as did copper. Aluminium could be used with good results, but in general gave inconsistent contacts and was too soft. Phosphor-bronze gave relatively poor results and tungsten also proved unsatisfactory.

Alloyed structures made by planar techniques are currently being developed. Such an example is illustrated in Fig. 5.7, with aluminium evaporated into a hole a few microns in diameter which is etched into a pyrolitically deposited layer of silicon dioxide on the surface of the n-type germanium, into which it is subsequently alloyed by a short heat treatment. Such structures may be formed as arrays on each germanium wafer and contacted by a method such as that indicated for Schottky-barrier diodes.

5.5 MOUNTING AND TESTING

The mounting and testing of backward diodes is generally effected as described for point-contact mixers and detectors, at least for those applications in which the backward diode is used as a direct replacement for a more conventional device. However, since the planar form does not necessarily require a catswhisker type of contact, it can be adapted so as to be suitable for strip-line.

REFERENCES

1 ENG, S. T., "Low noise properties of microwave backward diodes", *Trans. Inst. Radio Engrs*, **MMT-9**, p. 419 (1961).

2 WRIGHT, D. A., "Compound semiconductors", *Electron. Engng*, **31**, p. 659 (1959).

3 SPITZER, W. G., et al., "Properties of heavily doped n-type germanium", *J. Appl. Phys.*, **32**, p. 1822 (1961).

4 ESAKI, L., "New phenomenon in narrow germanium p–n junctions", *Phys. Rev. Letters*, **2**, p. 603 (1958).

5 OXLEY, T. H., and HILSDEN, F. J., "The performance of backward diodes as mixers and detectors at microwave frequencies", *Radio Electron. Engr*, **31**, p. 181 (1966).

6 SHURMER, H. V., "Backward diodes as microwave detectors", *Proc. Instn Elect. Engrs*, **111**, p. 1511 (1964).

7 OXLEY, T. H., "Backward diodes as mixers at microwave frequencies", *J. Electron. Control*, **17**, p. 1 (1964).

6

p-i-n Diodes

6.1 INTRODUCTION

p–i–n diodes (more strictly p^+–i–n^+) are used as variable-impedance elements to control r.f. signals, as attenuators, switches or modulators. They were developed in about 1950 as low-frequency rectifiers, but their

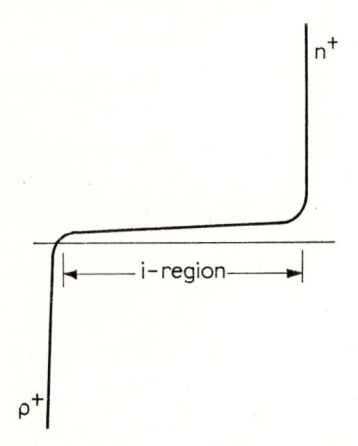

FIG 6.1 Doping profile of *p–i–n* diode

potential use at microwave frequencies first received wide publicity as a result of a paper by Uhlir [1].

The essential feature of such devices is a single *p–n* junction on one side of which is a relatively thick layer of high-resistivity silicon (the *i*-layer), a back contact being provided in the form of an additional thin layer of low-resistivity material of the same conductivity type. The end regions are highly doped and the transitions to them rapid, so that the impurity profile is of the form shown in Fig. 6.1. The manufacturing processes involve alloying, diffusion and epitaxial deposition, according to the application. Likewise the junction areas are controlled by any one of several available means, possibly covering an entire wafer or restricted to a mesa less than 100 μm in diameter.

At low frequencies the device behaves as a *p–n* junction rectifier, but at sufficiently high frequencies rectification ceases and the impedance becomes substantially that of the *i*-region. This will be high in the absence of bias or with the *p–n* junction biased in the reverse direction, but will become low when forward bias is applied to flood the *i*-region with injected carriers. It is clear that the frequency at which the device behaves in this way must be such that its period is small compared to the effective lifetime of the charge carriers. For most applications it is the two extreme states which are of interest, although the impedance can be varied continuously between these limits, a property utilized in variable attenuators. In the following discussion, the author has drawn freely upon a review article by J. E. Curran [2].

6.2 HIGH-IMPEDANCE STATE

6.2.1 Low Frequencies

At zero bias or under reverse voltage the depletion layer extends part-way into the *i*-region and the equivalent circuit is then as given in Fig. 6.2.

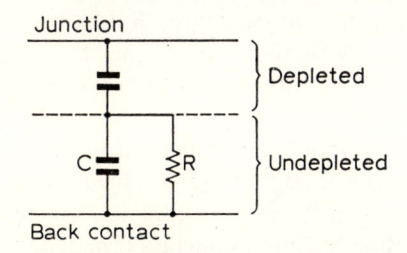

Fig 6.2 Equivalent circuit under reverse bias

The resistance of the depleted part is normally so high as to shunt the depletion capacitance to a negligible degree only.

The *Q*-factor of the undepleted part is given by

$$Q = \omega CR = \omega\tau \approx \frac{\rho}{5\lambda} \tag{6.1}$$

where ρ is the resistivity in ohm-centimetres, λ is the free-space wavelength in centimetres and τ is the dielectric relaxation time. For frequencies at which the Q-factor is less than unity ($c.$ 100 MHz) the undepleted part is primarily resistive and the diode behaves as a low-Q varactor. The impedance may nevertheless be sufficiently high for the non-linear effects to be tolerable. At low frequencies the peak voltage swing must not exceed the reverse bias voltage or carrier injection will occur, accompanied by a large drop in the impedance level.

As the reverse bias is increased, the width of the depletion layer increases until eventually it extends across the entire *i*-region. For reverse voltages exceeding this value the equivalent circuit is therefore that of a fixed capacitance shunted only by the very high resistance of the depleted region.

Capacitance is usually measured at about 1 MHz with a reverse bias of some 50 V, at which it may be assumed that the diode is fully depleted.

It is customary to quote a breakdown voltage also, but care must be taken in interpreting this. If the quoted value represents the onset of avalanche breakdown, the diode will have a short life if operated with a direct bias which approaches this value, an effect associated with the gradual migration of charge carriers and sharply accelerated at elevated temperatures. Thus the apparent voltage capability of the diode may not be realizable in practice and it may be necessary to limit the bias to as little as half this value.

The quoted breakdown voltage may sometimes represent a surface leakage limit, in which case it will be possible to approach it more closely, but a useful rule is to restrict the bias to about three times the value at which the capacitance is measured and to consult the manufacturers before planning to exceed this value.

Even with modest signal levels, sufficient bias to prevent forward drive may not be available, and it can then be helpful to put a high impedance in the bias path so that some self-bias may be obtained. Diode impedance is somewhat reduced in this situation but useful operation may none the less be possible.

6.2.2 High Frequencies

At frequencies which are well above 100 MHz the depletion boundary exhibits little response to the signal voltage, owing to the relatively long dielectric relaxation time associated with the high-resistivity silicon used in *p–i–n* diodes. Moreover, since the undepleted part has an impedance

p–i–n Diodes

which is primarily capacitive, there is only a very small overall variation in capacitance, whose effective value is that associated with the whole of the *i*-region and is independent of reverse bias voltage.

Surface perfection in terms of reverse leakage is much less important for *p–i–n* diodes than for most other devices, and a relatively large leakage has only a small effect on h.f. performance. Clearly it is important that any such effect should be stable or sufficiently small not to embarrass the bias supply circuit.

At frequencies above 1 GHz a high resistivity can be maintained for peak voltages which are an order of magnitude greater than the bias voltage. For example, a reverse bias of 100 V will enable a microwave voltage of 1 kV to be handled by a diode 0·23 mm in thickness. The reason why such forward drive produces only a slow fall in resistivity is that carriers do not have time between cycles to enter and accumulate in the *i*-region in sufficiently large numbers to produce a rapid degradation in resistivity.

The microwave voltage is limited on account of the above effect only by the permissible degradation in resistivity, and the process is reversible. The microwave stress can in fact exceed the d.c. breakdown stress of some 10 kV/mm, but a high breakdown voltage does not guarantee good microwave performance under these conditions.

6.3 LOW-IMPEDANCE STATE

When the diode is biased in the forward direction electrons from the n^+-side and holes from the p^+-side enter the *i*-region. These carriers set up a neutral plasma which is substantially uniform across the *i*-region, provided its thickness does not greatly exceed the diffusion length $\sqrt{(D\tau_c)}$, where D is the diffusion constant and τ_c is the carrier lifetime. The charge stored in this plasma is an increasing function of both thickness and current density, but the individual carriers, of course, eventually recombine.

Carrier recombination can occur either in the *i*-region or in the end regions, and the form of relationship between current density and diode resistance depends on which type of recombination predominates. High densities lead to more recombination in the end regions.

For a device having approximately parallel plane geometry and assuming a fairly uniform concentration of carriers in the *i*-layer, a general relationship for current density may be derived as follows. Provided that the *i*-layer thickness w is less than the ambipolar diffusion length L^*, it is very nearly true that for the densities of injected carriers in the *i*-region, particularly at high injection levels,

$$n = p = \text{a constant} = n_I$$

The first component of current density is that due to recombination in the *i*-layer and is given by

$$\frac{en_I w}{\tau_c} = \frac{en_I D^* w}{L^{*2}} \tag{6.2}$$

where τ_c = Minority carrier lifetime
D^* = Ambipolar diffusion coefficient = $2D_n D_p/(D_n + D_p)$

The suffixes *n* and *p* indicate the diffusion coefficients for electrons and holes.

Assuming that the quasi-Fermi levels do not change across the *p–i* or *i–n* junctions,

$$(pn)_P = (pn)_I = (pn)_N \tag{6.3}$$

where $(pn)_{P,N}$ refer to carrier concentrations just inside the *p*- and *n*-layers, respectively. Since the *p*-layer is highly doped, it follows that the concentration of holes just inside it is equal to the concentration of acceptor atoms, i.e. $p_P = N_A$. Similarly $n_N = N_D$ and also $p_I = n_I$. Thus

$$n_P \approx \frac{n_I^2}{N_A} \quad \text{and} \quad p_N \approx \frac{n_I^2}{N_D} \tag{6.4}$$

The electron current density flowing into the *p*-layer is the Shockley diffusion current:

$$J_{nP} = \frac{eD_n}{L_n} n_P \approx \frac{eD_n}{L_n N_A} n_I^2 \tag{6.5}$$

Similarly

$$J_{pN} \approx \frac{eD_p}{L_p N_D} n_I^2 \tag{6.6}$$

where D_n and D_p represent the diffusion coefficients for electrons and holes; similarly for the diffusion lengths L_n, L_p. The general relationship for current density across the junction is given by the sum of the components indicated in eqns. (6.2), (6.5) and (6.6), i.e.

$$J = \frac{eD_n}{L_n N_A} n_I^2 + \frac{eD_p}{L_p N_D} n_I^2 + \frac{eD^* w}{L^{*2}} n_I \tag{6.7}$$

p–i–n Diodes

For low current densities the third term in this equation predominates (mainly *i*-region recombination), and this leads to a relationship for the *i*-region low-resistance value which is of the form

$$R_l \propto \frac{w^2}{I_f \tau_c} \tag{6.8}$$

At high current densities the first two terms become the more important, leading to

$$R_l \propto \frac{w}{\sqrt{I_f}} \tag{6.9}$$

where I_f is the forward bias current. Thus the resistivity becomes independent of thickness and falls less rapidly with bias current. The transition between these two régimes usually occurs well before the maximum current is reached, at a current density of a few amperes per square centimetre. For such conditions the maximum permissible thickness is governed only by diffusion length, and in practice the diode thickness may be as much as 0·4 mm, before the effective resistivity begins to rise.

If we consider a more modest thickness of 0·15 mm, then a diameter of 0·4 mm can be tolerated before the capacitance exceeds 0·1 pF. Devices suitable for microwave use may therefore be made with relative ease and cheapness. The application of metal contacts directly to the semiconductor surfaces ensures that the resistance outside the *i*-region is negligible.

6.3.1 Resistivity under Forward Bias

The resistivity can be reduced to around 0·02 Ω-cm at a bias current density of some 3 000 A/cm² or about 0·15 Ω-cm at 50 A/cm². The drain on bias supplies is a prime factor limiting the current which can be used, and the former value is acceptable only where the active volume is small. At a current density of 50 A/cm² the 0·4 mm diode would require some 60 mA bias and would have a resistance of between 1 and 2 Ω, which is within the usually preferred limits of 0·5 and 3·0 Ω.

The microwave value of resistance under forward bias does not generally differ much from the d.c. slope resistance at the appropriate bias. This arises because, although most of the direct voltage drop may take place across the *i*-region, the slope resistance of the junction region can be much less than that of the *i*-region. Skin effect begins to increase the effective resistance if the diameter exceeds 1 mm at S-band or about 0·5 mm at X-band.

At high frequencies it has been found that the effective resistance begins to rise when the charge fluctuation (I_{rf}/w) is quite a small fraction of the total stored charge. Dr. Curran has reported that, with short pulses at S-band, currents greater than 10^4 A/cm^2 can be handled [2]. This is several hundred times the bias current density, the actual currents being well in excess of 100 A. The permissible peak current is found to fall with increasing frequency, but operation down to a few megahertz has been found possible with peak r.f. currents several times greater than the bias current.

6.4 SWITCHING RATIO

At zero bias a resistivity of at least 1 000 Ω-cm is readily obtainable, and this, in conjunction with a forward-biased resistivity of 0·2 Ω-cm, implies a switching ratio of at least 5 000:1. In general, however, the switching performance depends on the circuit conditions. At low frequencies the diode is usually operated with sufficient reverse bias to deplete the *i*-region completely, with the capacitance untuned. The ratio of this capacitive reactance to the forward-biased low resistance R_l is then $1/\omega C R_l$.

The product CR_l is usually about equal to the biased resistivity in ohm-centimetres when C is in picofarads and R_l in ohms. The capsule capacitance is usually approximately equal to that of the diode, so that for X-band, say, the above ratio will be only about 40 if the capacitance is not tuned out. For high-frequency applications it is usual to effect such tuning, the switching ratio then being limited only by the resistivity in the high-impedance state. The other principal consideration is that of bandwidth, implicit in the expression for Q-factor obtained by slightly modifying eqn. (6.1):

$$Q \approx \frac{\rho}{5\lambda}(1 + K) \qquad (6.10)$$

K being the ratio of capsule to diode capacitance.

The bandwidth implied by eqn. (6.10) is about $(1 + K) \times 5\%$ at X-band, for material of 1 000 Ω-cm resistivity. In a fully depleted diode the effective resistivity can be much higher than this, but the greater switching ratio which this suggests may not be realizable in practice owing to the correspondingly narrower bandwidth and the limitations of external losses.

In order to get minimum impedance in the forward-biased state it is desirable to tune out the diode inductance by inserting a capacitance in series with it. Such arrangements generally have a much lower Q-factor than is associated with the high-impedance state and do not seriously affect bandwidth. The diode inductance is almost entirely determined by the mounting arrangement rather than the semiconductor element—e.g. by the leakage path around pill-type diodes in low-height waveguide, or by the length of wire or tapes for diodes terminated in this way which are used in coaxial—or strip—transmission lines.

6.5 SWITCHING SPEED

The switching time between extreme states depends very much on the bias circuit conditions. A relatively high density of carriers is required for the attainment of low resistivity under forward bias. This charge has to be removed almost completely before the high-impedance state is regained, and the final stage of removal involves a complicated voltage build-up, which is difficult to predict. However, it is usual to specify the switching time as that between 10% and 90% transmission when the diode is situated in series with a 50 Ω transmission line, and it is relatively simple to calculate this time for the conditions which usually apply.

Four distinct situations are possible, each leading to a different solution. Two of these pertain to the sites at which recombination of the charge carriers occurs, whether these are predominantly in the p^+ or n^+ end regions or in the i-region. The other governing factor is whether or not the thickness of the latter is large or small compared to the carrier diffusion length. p–i–n diodes usually involve recombination mainly in the i-region, and their thickness is made less than a diffusion length—otherwise the forward resistance rises rapidly. For these conditions the stored charge Q_s required for the establishment of the low-resistivity state is governed by relationship

$$Q_s = I_f \tau_c \qquad (6.11)$$

where I_f is the forward bias current. It follows that, with a constant applied bias current, a fraction $(1 - 1/e)Q_s$ is injected in a time during which the resistance falls to $1 \cdot 5 R_l$, R_l being the final low-resistance value. Thus, for fast switching into the low-resistance state, a short carrier lifetime is required. On this basis it may be shown that a diode with an i-region thickness of $10\,\mu$m at a bias current density of $50\,\mathrm{A/cm^2}$ will attain a resistivity of $0 \cdot 2\,\Omega$-cm if the lifetime does not exceed 5×10^{-8} s.

Microwave Semiconductor Devices

If there is sufficient switching voltage available to clear all the stored charge, the turn-off time t_0 is given approximately by

$$t_0 = \frac{Q_s}{I_f + I_r} = \frac{I_f \tau_c}{I_f + I_r} \tag{6.12}$$

where I_r is the available reverse current. Since, under pulsed conditions, I_r may be made large with respect to I_f, turn-off times may be much less than the carrier lifetime. An ultimate limit to switching speed is set by the transit time through the i-region, but even for a diode of 0·25 mm thickness this is only about $2·5 \times 10^{-9}$ s. In practice such diodes can be made to switch in about 2×10^{-8} s.

For a given biased resistance, the stored charge is proportional to the square of the i-region thickness, so that thinner diodes can be switched faster and with less reverse bias current, but they are limited in power-handling capability. The peak reverse voltage is limited by the thickness of the diode on account of breakdown, which occurs at a stress of about 10^5 V/cm, the voltage corresponding to a thickness of 10μm thus being 100 V. The peak current also is limited on account of the smaller value of stored charge (rectification may occur at low frequencies). Thin diodes have the further disadvantage that for a given impedance the area is smaller, leading to higher thermal resistance.

For fast switching, the bias circuit must be able to provide adequate bias current or voltage rapidly. The high-impedance state imposes the more severe conditions, since almost the whole of the stored charge must be removed. The final stage of the turn-on phase is not so important, however, and provided the effective carrier lifetime is short enough, it is not necessary to apply a "forcing" waveform when switching into forward conduction.

6.6 POWER CAPABILITY

The mean power is usually limited by thermal resistance, and the peak power by voltage or current, according to the application. Thermal resistance can be as low as 10°C/W or even less but is more usually around 20–30°C/W. Dissipation in the diode may be up to 10 W, with a corresponding mean power-handling capability up to 100 W. Much depends on how good the contact arrangements can be made in relation to thermal dissipation.

6.7 APPLICATIONS

6.7.1 General

The principal methods of using microwave p–i–n diodes can be identified with four types of application, as follows.

Square-wave Modulators

With square-wave modulators it is necessary only to interrupt the flow of power along a transmission line for successive time intervals, and it is not of prime importance what happens to the power which is stopped. Here it is quite acceptable for power to be absorbed in the diode (subject to allowable dissipation) or in some other resistive element, suitably coupled.

Switches

With a switching device, it is necessary to be able to transmit the microwave power along one or other of alternative routes with minimum attenuation in each case. This is accomplished by arranging that most of the power is transmitted past the diode in the one state and reflected from it in the other. No deliberate absorption of power in either state is permissible.

Phase Shifters

The diode is here usually situated in a branch arm coupled to the main transmission line, the phase of the reflected power being dependent on the bias applied to the diode. A common arrangement is to terminate the branch arm is a short-circuit with the diode in front, when incoming power is reflected either mainly by the diode or mainly by the short-circuit, the phase difference being determined by their electrical separation.

Voltage-controlled Attenuators

As variable attenuators, diodes are required which give specified laws of attenuation versus forward bias, so that the attenuation may be pre-set by the bias or programme controlled. Alternatively, a modulation envelope of prescribed waveform may be imposed on the microwave signal.

Power Limiters

Active or passive limiters can be made using p–i–n diodes. Pre-pulsed units are being used in duplexing, and Muskus [3] has described the incorporation of standard diodes both within and external to a T-R cell for limiting

Microwave Semiconductor Devices

the breakthrough "spike" which is transmitted to a mixer. (*Note*. A *pre-pulsed limiter* is one triggered from the modulator which fires the magnetron, so that the limiter is already in its attenuating state when the "spike" arrives.)

More recent work has been concerned with the development of *p–i–n* diodes as active limiters, but there is as yet little detailed information available on these.

6.7.2 Square-wave Modulators

The simplest type of modulator in common use consists of a diode shunting a matched transmission line, as shown in Fig. 6.3. If the admittance of

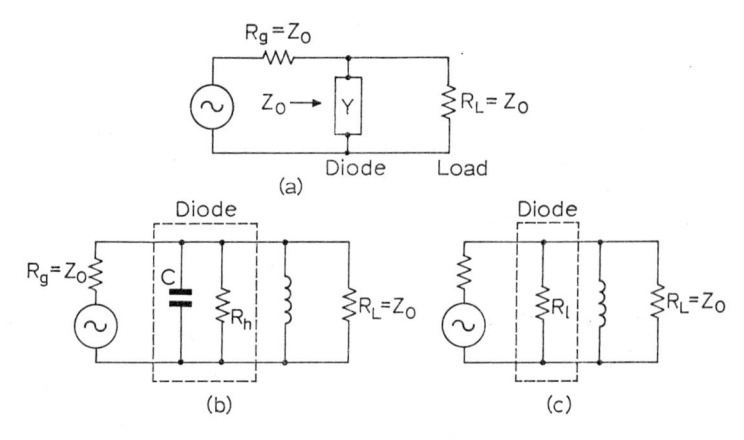

FIG 6.3 Simple modulator equivalent circuits
    (*a*) General circuit
    (*b*) Zero bias
    (*c*) Forward bias

the diode normalized to that of the transmission line is represented by $y \equiv g + jb$, it is easily shown that the insertion loss L_i is given by

$$L_i = (1 + g/2)^2 + (b/2)^2 \qquad (6.13)$$

It is usual to tune out the capacitive reactance of the diode in the high-impedance state by means of a shunt inductance, which may generally be ignored in the forward-biased condition. Using the suffixes *h* and *l*

p–i–n Diodes

to indicate the high- and low-impedance states, we obtain from eqn. (6.13), at the centre frequency,

$$\left. \begin{array}{l} L_h = \left(1 + \dfrac{Z_0}{2R_h}\right)^2 \\[2mm] L_l = \left(1 + \dfrac{Z_0}{2R_l}\right)^2 \end{array} \right\} \qquad (6.14)$$

where Z_0 is the characteristic impedance of the transmission line.

We have previously noted that the effective diode resistivity can be switched between, say, 1 000 Ω-cm and 0·2 Ω-cm. This switching range may be made to correspond to the resistance values appropriate to modulation in a feeder of given characteristic impedance, e.g. $R_h = 5\,000\,\Omega$, $R_l = 1\,\Omega$ with $Z_0 = 250\,\Omega$. The corresponding values of attenuation are $L_h = 0\cdot2\,\mathrm{dB}$, $L_l = 42\,\mathrm{dB}$.

The useful bandwidth for a modulator of the above type is determined by the susceptance B of the diode capacitance in parallel with its tuning inductance, L. The susceptance at any frequency ω is given by

$$B = \omega C - \frac{1}{\omega L} \qquad (6.15)$$

If this circuit is resonant at an angular frequency ω_0, then, since $\omega_0 C = 1/\omega_0 L$, provided that ω is not very different from ω_0, we may write

$$B = C\Delta\omega$$

where $\Delta\omega = 2(\omega_0 - \omega)$.

If the bandwidth $\Delta\omega$ is defined as the frequency range over which the insertion loss due to the uncompensated susceptance is less than, say, 0·5 dB (i.e. a factor of 1·12) at the edges of this band, it follows that

$$1\cdot12 = \left(1 + \frac{Z_0 C \Delta\omega}{2}\right)^2 \qquad (6.16)$$

For a frequency of 10 GHz, with $Z_0 = 250\,\Omega$ and $C = 0\cdot2\,\mathrm{pF}$, the fractional bandwidth on this basis is 0·038. A wide bandwidth clearly requires low capacitance and a low-impedance feeder.

One arrangement utilizing both of the above principles has been described by Baker [4], in which the parallel capacitance was minimized by using diode elements mounted unencapsulated on a brass plunger ("stick"

Microwave Semiconductor Devices

diode) and inserted across a section of waveguide tapered to a point of low impedance, as illustrated in Fig. 6.4. Using early diode material, such an arrangement gave an attenutation of 20 dB with an insertion loss of less than 2 dB over the waveband 3·0–3·5 cm.

Much greater bandwidths have been achieved using "stick" diodes arranged in multiple configurations of two or three diodes, as described by Pavey [5]. In these arrangements local tuning in the form of a small capacitance in series with each diode is used to minimize the effects of series inductance on attenuation. A typical performance achieved for this

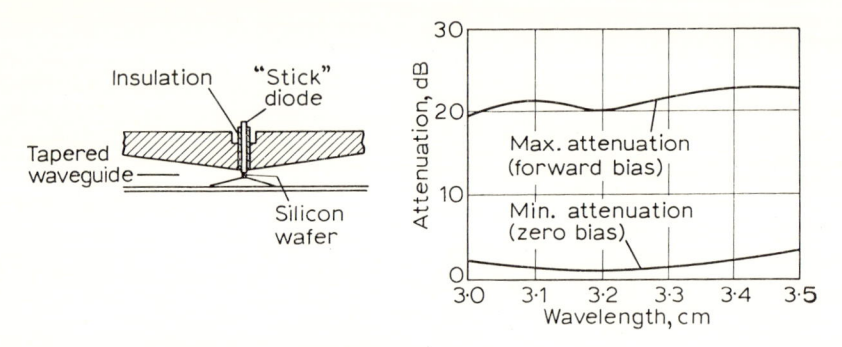

Fig 6.4 Broad-band single diode mount
(From Ref. 4)

type of modulator is an attenuation range of 1–20 dB over the frequency band 8–16 GHz, using two diodes.

High switching ratios have been achieved in balanced systems in which the reflected power is cancelled out in the high-attenuation state. Anderson [6] has described two-diode arrangements capable of switching 20 W of X-band power with an insertion loss of less than 0·5 dB and an attenuation of 50 dB, at switching speeds in the region of about 0·5 μs. Two types of system of this kind are illustrated schematically in Fig. 6.5. Both are narrow-bandwidth systems, the narrow-wall coupler being the more restricted on this score but capable of higher switching ratios for a given diode specification.

Roberts and Robinson [7] have described a very broadband modulator employing a coupled-line hybrid circuit. With their arrangement the incident power in the "isolated" condition is absorbed in the p–i–n diodes, which are matched into two arms of the coupler. The bandwidth achieved for attenuation between 2 dB and about 15 dB is 2·5–7·5 GHz.

p–i–n Diodes

Further applications of p–i–n diodes are described in the Proceedings of the Joint IEE/IERE Symposium on Microwave Applications of Semiconductors, University College, London, July, 1965. More recent work has

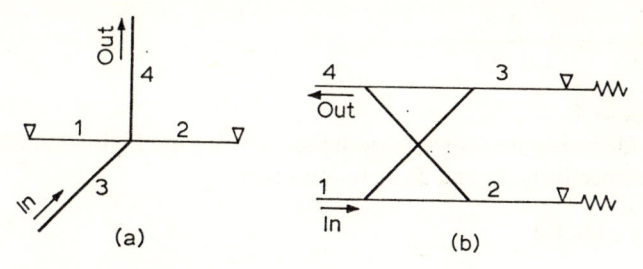

FIG 6.5 (a) Magic-T switch, (b) narrow-wall coupler switch
(*From Ref. 6*)

been directed towards incorporating p–i–n diodes in hybrid and monolithic integrated circuits.

6.7.3 Switches

The basic situation relating to a diode intended for switching microwave power between two loads is illustrated in Fig. 6.6(a). When the diode

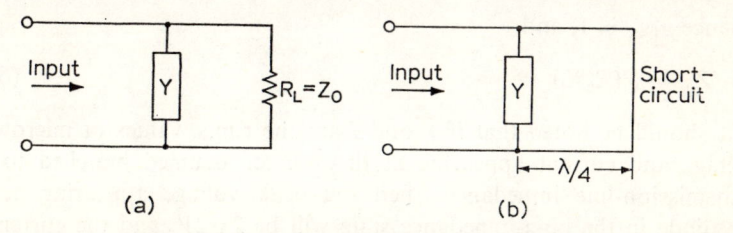

FIG 6.6 (a) Basic switch circuit, (b) basic 180° phase-shift circuit

conductance is low, most of the incoming power is transmitted past the diode into the load R_L, assumed matched to the feeder impedance Z_0. With the diode switched into the high-conductance state, most of the incident power is reflected and it is assumed that a coupling arrangement (not shown) is available on the input side which will divert this power into a second load which is also matched to the line. Usually it is required that the line impedance Z_0 be chosen so that the power loss due to the finite switching range of the diode shall be equalized between the two states.

Microwave Semiconductor Devices

If the normalized conductance of the diode in its low-impedance state is g_1 and in the high-impedance state g_2, equating the power reflected in the former state to the power absorbed in the load in the latter gives

$$\frac{g_1^2}{(2+g_1)^2} = \frac{4}{(2+g_2)^2} \tag{6.17}$$

whence $g_1 g_2 = 4$.

If the diode resistances corresponding to the low- and high-conductance states are respectively R_l and R_h it follows that

$$Z_0 = 2\sqrt{(R_h R_l)} \tag{6.18}$$

6.7.4 Phase Shifters

The use of *p–i–n* diodes in phase shifting is illustrated in Fig. 6.5(*b*) for an arrangement introducing a phase shift of 180°.

To equalize the losses in this case, the reflection coefficients of the combination of diode and short-circuit termination must be equal in magnitude but opposite in phase for the two states of the diode, i.e.

$$\frac{g_1 - 1}{g_1 + 1} = -\frac{g_2 - 1}{g_2 + 1} \tag{6.19}$$

whence $g_1 g_2 = 1$, and

$$Z_0 = \sqrt{(R_h R_l)} \tag{6.20}$$

It should be noted that if V and I are the r.m.s. values of microwave voltage and current appearing at the source, assumed matched to the transmission-line impedance, then the peak voltage appearing across the diode in the high-impedance state will be $2\sqrt{2}V$, and the current in the low impedance state $2\sqrt{2}I$.

6.7.5 Voltage-controlled Attenuators

With voltage-controlled attenuators the variation of diode conductance with forward current is of prime importance. However, because of the existence of two distinct régimes of operation under forward bias, the range between the extreme bias states is rather difficult to predict. Nevertheless, by the appropriate selection of diodes, arrangements have been devised in which adequate control has been achieved for specific applications.

p–i–n Diodes

Allen [8] has described such circuits at X-band based on a magic-T system, using coaxial diodes, and James and Potok [9] have discussed a broadband unit designed for S-band. Hewlett–Packard [10] have described a symmetrical π-network for lower frequencies which gives a constant-impedance controlled attenuator over the bandwidth from 10 MHz to 1 GHz.

REFERENCES

1 UHLIR, A., "The potential of semiconductor diodes in high-frequency communications", *Proc. Inst. Radio Engrs*, **46**, p. 1099 (1958).
2 CURRAN, J. E., "High-frequency p–i–n diodes", *Electron. Equipment News* (Nov., 1967).
3 MUSKUS, C. Z., "Some applications of the microwave diode switch", *Digest of I.E.E. Conference on Components for Microwave Circuits*, p. 55 (Sept., 1962).
4 BAKER, T. H. B., "Semiconductor diode waveguide switch," *Electron. Tech.*, p. 300 (Aug., 1961).
5 PAVEY, N. A. D., "Broad-band p–i–n diode switches for the 8–16 GHz range" *Proc. IERE/IEE Joint Symposium on Microwave Applications of Semiconductors* Paper 6 (July, 1965).
6 ANDERSON, D. W., "Microwave switching with variable resistance diodes", *Digest of IEE Conference on Components for Microwave Circuits*, p. 55 (Sept., 1962).
7 ROBERTS, D. A. E., and ROBINSON, S. J., "A p–i–n diode modulator for the frequency band 2·5–7·5 GHz", *ibid.*, p. 37.
8 ALLEN, S. G., "Microwave variable attenuator using p–i–n junction diodes", *Brit. Comm. Electron.*, p. 123 (1963).
9 JAMES, J. R., and POTOK, M. H. N., "The design of p–i–n diode coaxial attenuators and levellers", *Electron. Engng*, p. 142 (1966).
10 HEWLETT PACKARD ASSOCIATES, *Application Note 912—An Attenuator Design using p–i–n Diodes.*

7

Transistors

7.1 INTRODUCTION

Essentially, a transistor consists of two rectifier-type junctions back to back, a basic arrangement being shown at (*a*) in Fig. 7.1, with the corresponding potential distribution illustrated at (*b*). In this example the base region is

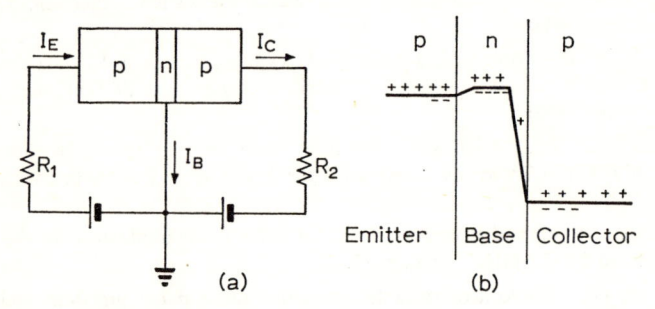

FIG 7.1 Principle of action of *p–n–p* transistor
  (*a*) Basic arrangement
  (*b*) Potential diagram

of *n*-type material, the emitter *p–n* junction being positively biased and the collector *n–p* junction negatively biased.

The majority carriers in the base region are electrons, which are repelled by the bias across the collector junction. The forward bias applied to the emitter junction, however, leads to the injection of minority carriers (holes) into the base and these carriers are attracted to the collector when they fall under the influence of the field in the vicinity of the *n–p* junction. In order to facilitate the survival of minority carriers long enough to drift across the base region, the latter should be very thin, a feature which is also desirable in order that the collector current may respond rapidly to changes in emitter current. It is usual for some 2% of the injected holes to recombine in the base, thereby giving rise to a base current which is about 2% of the emitter current.

The amplifying properties of a transistor derive from the fact that the collector behaves as a generator of high resistance compared with that of the emitter, since the collector current is substantially independent of collector voltage over a wide range of bias. As the collector current is only slightly less than the emitter current and is passed through a much higher resistance, there is power gain between the emitter and collector circuits, which can be substantial.

If a signal is applied between the base and emitter with output taken between the collector and emitter (common-emitter arrangement) current gain results. This is usually indicated by the symbol β for an output short-circuited to alternating current, typical currents gains being well in excess of 100 times. For a common-base arrangement the corresponding factor, denoted by α, is less than unity and is equal to $\beta/(1 + \beta)$.

The cut-off frequency f_c is taken as that at which the current gain has fallen to $1/\sqrt{2}$ times its low-frequency value. With a common-emitter circuit this is much less than for a common-base arrangement, being approximately $1/\beta$ times the value appropriate to the latter case.

The earliest transistors, made in 1948 at the Bell Telephone Laboratories [1], had a point-contact emitter and collector. Subsequent history has been excellently reviewed by Tenzer [2], up to 1961. The first major milestone in device development was the grown junction transistor. Such devices were made by crystal growing techniques, all involving initially the dipping of a seed crystal into molten germanium or silicon doped either p- or n-type. After growing sufficient material of the one kind, the opposite type of dopant is added and so on until a final crystal is obtained which can be cut into many transistor bars. Transistors of this kind, however, tend to have high base and series collector resistance.

Alloy junction types were the first devices capable both of high power dissipation and moderately fast switching. Such transistors employ crystals of one impurity type only and are uniformly doped, the emitter and collector junctions being formed by fusing into the base material pellets containing appropriate doping agents of opposite conductivity type. An example of this type of transistor is illustrated in Fig. 7.2(a). The upper frequency limit is largely determined by the thickness of the base region, which must be adequate to withstand the required collector voltage.

By incorporating a layer of intrinsic semiconductor material between the base and collector it is possible both to reduce the output capacitance

Microwave Semiconductor Devices

and increase the output power, as in the *p–n–i–p* transistor, shown in Fig. 7.2(*b*). Another way of reducing the collector capacitance as well as the base resistance is employed in the drift transistor, in which an impurity gradient is produced by diffusing additional doping agents into the base material. Such devices have α-cut-off frequencies up to about 100 MHz.

The surface-barrier transistor (SBT) has similar frequency limitations and is illustrated in Fig. 7.2(*c*). This type of device is made by chemical

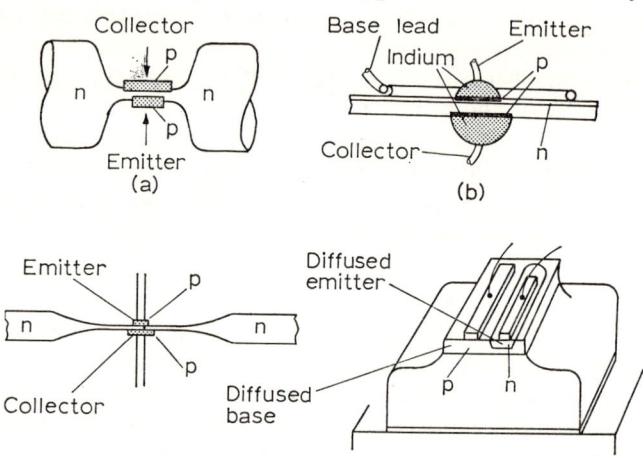

FIG 7.2 Construction of various h.f. transistors
(a) Alloy junction
(b) *p–n–i–p*
(c) Surface barrier
(d) Conventional mesa
(*From Ref.* 2)

etching techniques in which the semiconductor material is used as the anode and the nozzle of a very fine jet of etching fluid is the cathode. After the desired base thickness has been achieved the emitter and collector are formed through plating into the etched cavities.

Next in chronological order is the micro-alloy transistor (MADT), exhibiting α-cut-off frequencies up to 1 GHz and capable of oscillation in the 10 cm band. This device employs diffusion into the base, as with the drift transistor, but the emitter is formed by alloying.

Mesa transistors followed, in which base and emitter layers are formed on a collector substrate. These employ processing which includes masking techniques, diffusion, evaporation and etching. Gold-alloy stripes are evaporated and alloyed over the base and emitter regions to provide ohmic contacts. Such a device is illustrated in Fig. 7.2(*d*).

Transistors

The advent of epitaxial techniques led to transistors of the form illustrated in Fig. 7.3, which relates to an example developed at the Bell Telephone Laboratories. Tenzer, in his review paper [2], quoted this device as having a mesa diameter of 0·04 mm with base and emitter stripes each about 0·08 mm wide. Compression-bonded gold leads are attached to each of them. Using epitaxial techniques, base and collector depletion-layer thicknesses can be accurately controlled to less than 0·1 μm. Planar

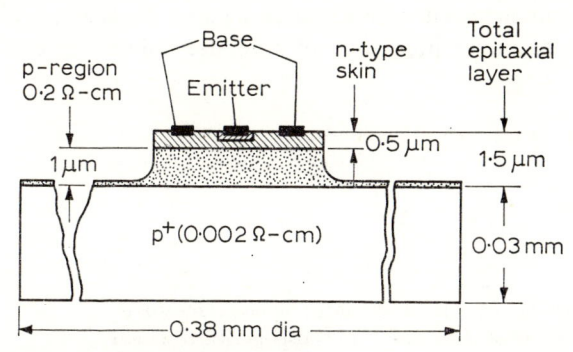

FIG 7.3 Epitaxial mesa transistor
(From Ref. 2)

techniques were subsequently developed, using similar masking and etching stages but with a relatively thick passivating layer of silicon oxide retained over the exposed substrate surfaces.

All the above devices employ charge carriers of both sorts (electrons and holes), and are known as *bipolar* transistors, of which the *n–p–n* type in silicon is probably the most versatile. Field-effect transistors (FETS) operate with carriers of only one sort and are therefore said to be *unipolar*. There are two types of FET: the junction-gate field-effect transistor (JUGFET), and the insulated-gate type (IGFET), often called the metal-oxide-semiconductor transistor (MOST).

7.2 DESIGN CONSIDERATIONS FOR POWER DEVICES

7.2.1 Bipolar Transistors

For high-frequency operation the transit time τ_{ec} between emitter and collector must be small. We have the relationship

$$\tau_{ec} = \frac{1}{2\pi f_T} \tag{7.1}$$

Microwave Semiconductor Devices

where f_T is the frequency at which $\beta = 1$. In order that τ_{ec} shall be sufficiently small the base width w must be made less than $1\,\mu\text{m}$. A typical high-frequency equivalent circuit is shown in Fig. 7.4.

The base current which flows under the emitter region gives rise to a transverse bias field which results in constriction of current emanating from the emitter junction. For maximum current and hence maximum power output the emitter periphery should therefore be as long as possible. However, this requirement must be balanced against the need to restrict the areas of the base and emitter, so as to minimize emitter and collector

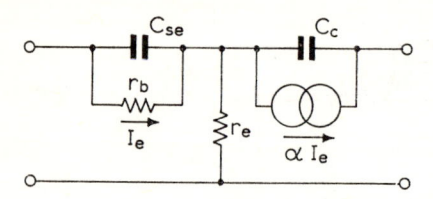

FIG 7.4 Typical h.f. equivalent circuit for bipolar transistor

r_b = base spreading resistance C_{se} = emitter storage capacitance
r_e = emitter resistance C_c = collector-to-base capacitance

transition-region capacitances, both of which tend to increase the delay time. Structures most effectively satisfying these requirements can be made with narrow-stripe interdigited emitter and base configurations.

Lathrop [3] has derived a figure of merit f_{max} for such a structure as

$$f_{max} = \frac{40}{s + 2t} \text{ gigahertz} \qquad (7.2)$$

where the stripe width s and the spacing t are both in microns. Thus for f_{max} to be in the 3 cm band both s and t would have to be of the order of $1\,\mu\text{m}$, which can just be achieved under laboratory conditions.

Pritchard [4] has described the combined effects on f_{max} of collector capacitance C_c and base resistance r_b in terms of the product $r_b C_c$, which is independent of the stripe length l, provided that the emitter-edge current density is maintained constant. This arises because C_c is directly proportional to l, whereas r_b is inversely proportional to it. Similarly it may be shown that other resistance-capacitance products affecting τ_{ec}, i.e. $r_{sc}C_c$ and $r_e C_{Te}$, are also independent of l, where r_{sc}, r_e and C_{Te} are the collector series resistance, emitter resistance and transition capacitance, respectively. Thus the high-frequency performance is unimpaired by connecting together an arbitrary number of transistors as parallel elements, which provides a convenient means of obtaining high power.

Practical limits inevitably arise in relation to the number of elements which can be interconnected without significantly degrading the high-frequency performance or incurring thermal limitations. There are also fabrication problems in preparing fault-free epitaxial material and maintaining uniform processing over large areas.

With silicon transistors, the permissible junction temperature is about 200°C, roughly twice that for germanium transistors. Since the thermal conductivity of silicon is 2·54 times that of germanium, the former material is the obvious choice for power transistors. Satisfactory operation of large-area structures depends on the ease with which heat may be conducted away from the entire area, a problem which is particularly acute at low frequencies.

The form of encapsulation is clearly important from both electrical and thermal considerations. The upper frequency limit imposed by the container should considerably exceed the operating frequency, which implies that all interconnections and shunt capacitances should be kept small. Parasitic losses must be minimized, and at frequencies above about 1 GHz this necessitates the use of distributed elements rather than lumped circuitry. For use below 3 GHz waveguide is unsuitable and the mounting arrangement must be appropriate to coaxial-, strip- or trough-line transmission systems. For efficient conduction of heat away from the chip whilst maintaining electrical isolation, the latter may be mounted on a disc of beryllium-oxide ceramic.

Use may be made of the properties of impedance-transforming networks to overcome the problems of operating high-power transistors of some hundreds of watts with very low input impedances. By employing such a network the input impedance may be increased by a factor approximately equal to the square of the Q-factor of the network.

The *maximum frequency of oscillation*, a term which was introduced by Pritchard [4], is the frequency at which the power gain has dropped to unity. It may be written as

$$f_{max} = \left[\frac{f_T}{8\pi r_b C_c}\right]^{1/2} \approx 200 \left[\frac{\alpha_0 f_T}{r_b C_c}\right]^{1/2} \tag{7.3}$$

with f_T in megahertz and C_c in picofarads, and the power gain is given by

$$G = \left(\frac{f_{max}}{f}\right)^2 \tag{7.4}$$

Here α_0 indicates the low-frequency value of α, determined from the delay time, which depends on a variety of factors such as the current level, impurity distribution and diffusion constant in the base region, collector depletion layer width and minority carrier velocity. The detailed effect of these factors has been discussed by J. M. Early [5].

Factors tending to decrease f_T arise from various causes, e.g. decreased emitter efficiency at high current levels, which produce additional non-linear effects, and to increased temperature as a result of thermal dissipation. At high currents there may occur a phenomenon known as the Kirk effect [6], involving a widening of the effective base region. The latter may also indirectly result in restricted power output on account of field limitation of the collector voltage.

Regarding the product $r_b C_c$, conductivity modulation of the base region helps to keep r_b small, but because of non-uniformity in current at the emitter edge the effective collector capacitance is several times that which it would otherwise be, further contributions arising from the package and strays. It has been shown by E. O. Johnson [7] that

$$V_m f_T = \begin{cases} 2 \times 10^{11} \text{V/s for silicon} \\ 1 \times 10^{11} \text{V/s for germanium} \end{cases} \qquad (7.5)$$

where V_m is the maximum permissible collector voltage. In practice it is necessary to limit the applied collector voltage to less than $V_m/2$, since the peak value will be twice that of the supply. This leads to the expression for output power, assuming matched conditions,

$$P = \frac{V_m^2}{16 R_L} \qquad (7.6)$$

If the output resistance of the transistor is matched to the load resistance R_L, then

$$R_L = \frac{1}{2\pi f_T C_c} \qquad (7.7)$$

From eqns. (7.3), (7.5) and (7.7) we obtain

$$P_{max} f^2{}_{max} = \frac{6 \cdot 25 \times 10^{20}}{r_b} \text{ watt-hertz}^2 \text{ for silicon} \qquad (7.8a)$$

$$= \frac{1 \cdot 56 \times 20^{20}}{r_b} \text{ watt-hertz}^2 \text{ for germanium} \qquad (7.8b)$$

At any particular frequency, f, the output power, P, is given by

$$\frac{P}{P_{max}} = 1 - \left(\frac{f}{f_{max}}\right)^2 \tag{7.9}$$

Eliminating P_{max} in eqn. (7.8a) and differentiating we have, for silicon,

$$f = 0.7 f_{max} \quad \text{so that} \quad P = 0.5 P_{max}$$

Substituting these results in eqn. (7.8a),

$$Pf^2 = \frac{1.56 \times 10^{20}}{r_b} \text{ watt-hertz}^2 \tag{7.10}$$

The minimum value for r_b is about $0.5\,\Omega$, so that in practice the limiting performance of a silicon transistor as an oscillator is given by

$$Pf^2 = 300\,\text{W-Ghz}^2 \tag{7.11}$$

For germanium the corresponding performance figure is only about one-quarter of the above.

If it assumed that for a bipolar transistor $t = S/2$, then eqn. (7.2) may be written

$$f_{max} = \frac{20}{S} \text{ GHz} \tag{7.12}$$

In practice f_{max} is usually greater than f_T by a factor of two or three. Similar expressions have been derived by M. B. Das [8] for both types of field-effect transistor, and experience shows that the value of f_{max} attainable is about the same for bipolar and unipolar transistors.

7.2.2 Requirements for Low-noise Amplifiers

For amplifier applications low noise figure is an important requirement additional to that of power gain. The frequency f_α at which α has decreased to $\alpha_0/\sqrt{2}$ should be as high as possible to ensure that uncorrelated collector noise is a minimum [9]. Any excess base resistance will here act as a source of thermal noise as well as imposing a limit on the power gain. Germanium transistors are often superior to silicon ones in these applications, mainly because the higher mobility of germanium leads to lower values of r_b.

In order that the shot noise should be low, a high current gain must be achieved at a fairly low current, which implies that C_{Te} should be low.

Microwave Semiconductor Devices

This is aided by keeping the emitter area small and the base resistivity high, and clearly a compromise has to be made between r_b and C_{Te}.

At frequencies below about 1 GHz the noise figure of a transistor amplifier is competitive with that of a parametric amplifier and the circuitry is simpler.

An expression for the noise figure which has been found to agree closely with measured values is [10, 11]

$$F = 1 + \frac{r_b + r_{e/2}}{R_g} + \frac{(r_b + r_e + R_g)^2}{2\alpha_0 R_g r_e} \left[\frac{1}{h_{FE}} + \left(\frac{f}{f_\alpha}\right)^2 + \frac{I_{CBO}}{I_E} \right] \quad (7.13)$$

where h_{FE} = Common-emitter static forward-current transfer ratio with output voltage held constant, commonly known as the common-emitter d.c. current gain

R_g = Source resistance

I_{CBO} = Collector current for open-circuited emitter, commonly known as the common-base leakage current

I_E = Emitter current

A general discussion of microwave transistor design problems has been given in a series of articles by J. G. Tatum [12].

7.3 MANUFACTURE

Microwave transistors may be prepared by photo-lithographic processes associated with planar technology or fabricated using electron-beam techniques, the former being the more usual method. The formation of *p–n* junctions is usually achieved by diffusion, but ion implantation is an alternative. Base widths down to $0.2\,\mu$m are currently employed.

An arrangement suitable for transistors with f_{max} up to 10 GHz is illustrated in Fig. 7.5, which shows a series of interdigital base and emitter stripes, the collector being underneath. In practice as many as 100 emitters may be used, stripe widths being as narrow as 1–$2\,\mu$m.

Epitaxial material is widely used, the substrate resistivity being of the order of $0.01\,\Omega$-cm and the film about two orders higher and with a thickness up to $10\,\mu$m, resistivity and thickness being particularly critical in relation to power devices. Dislocation density should be less than $1000/\text{cm}^2$. The use of epitaxy aids in reducing the collector capacitance and also the spreading resistance. Reduction of this resistance decreases the output conductance and thus raises the transistor gain. Also, control of the

Transistors

resistivity of the collector adjacent to the base helps to maintain the value of current gain up to higher current levels.

7.4 APPLICATIONS

The utilization of transistors in the microwave field is steadily increasing. Although transistors supplying power directly at 10 GHz are not currently available, a relatively cheap transistor operating at 1–2 GHz and followed

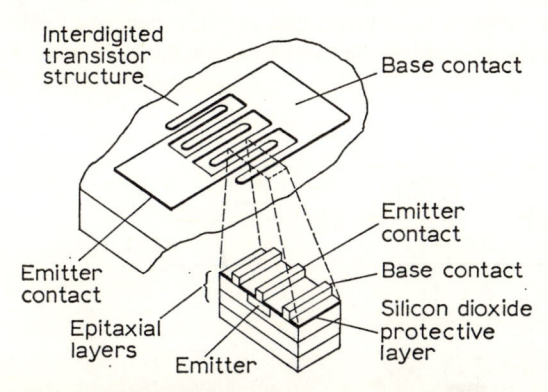

Fig 7.5 Interdigited planar transistor structure
(From Ref. 18)

by a step-recovery diode makes a very competitive solid-state source for the 3 cm band.

Two cavities are usually required by transistor microwave oscillators, for the emitter–base and collector–base circuits, the former controlling feedback and the latter the frequency of oscillation. At lower frequencies it is possible to replace the emitter–base cavity by a choke or even a resistor. We note that eqn. (7.11) predicts a limiting performance of about 75 W at 2 GHz for a silicon transistor, which compares with about 1 W currently achieved at that frequency and is indicative of the scope for development which remains.

Power may be generated above the normal cut-off frequency by simultaneously operating a transistor as a power amplifier and varactor multiplier [13].

The replacement of parametric amplifiers at frequencies of 1–2 GHz may be particularly important in relation to phased-array aerial systems.

Microwave Semiconductor Devices

Circuit Notes

The choice of circuit configuration depends in general upon several features—operating frequency, circuit function, transistor parameters and impedance levels. Common-base arrangements are characterized by low input and high output impedance and they also involve positive internal feedback due to the collector capacitance, which may lead to instability if the circuit is not loaded correctly. As the current gain is less than unity, power gain is achieved by impedance transformation. Common-emitter circuits have high input and lower output impedances, and produce negative feedback, which tends to reduce the gain unless the collector capacitance is neutralized.

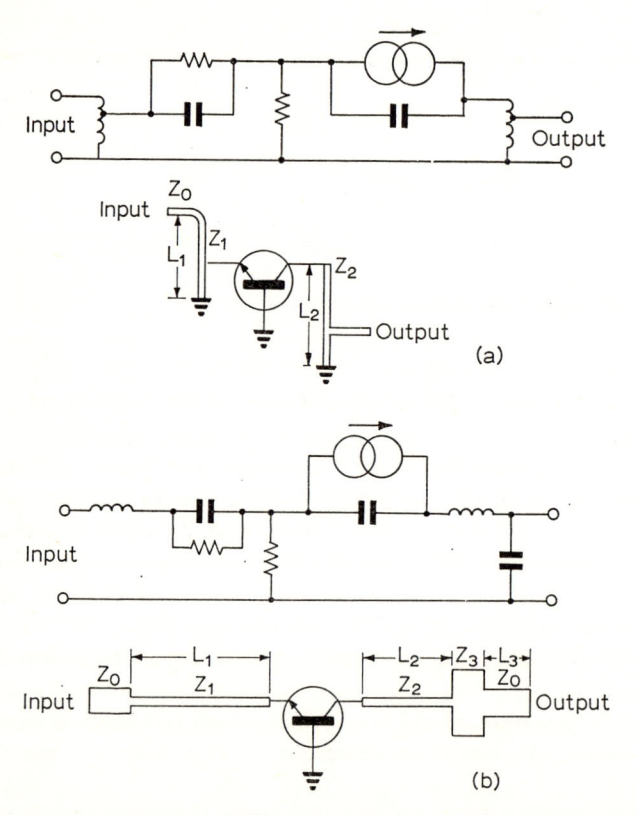

FIG 7.6 Amplifier equivalent circuits with transmission line analogue
(From Ref. 19)

Transistors

Provided that the operating frequency is less than $f_T/2$, this mode is usually preferred, except at the highest frequencies, where the common-base arrangement is used.

Figure 7.6 shows two possible arrangements (for common-base connection) with their transmission line analogues, which can give good gain-bandwidth products and low noise figures.

Balanced and unbalanced strip-line configurations are particularly appropriate forms of transmission line. Gelnovatch [14] has indicated that it is not necessary to employ a rigorous approach and has described a construction procedure which is said to be rapid, correlating well with predicted performance.

Varactor diodes may be used as transistor tuning elements. Typically the total shunt capacitance of the transistor will be about 2 pF, which is an order of magnitude greater than the minimum capacitance of otherwise appropriate microwave varactors. To enhance their effect, however, use may be made of an impedance transformation via a transmission line, as described by Herbert and Chernega [15], who indicated as an example an octave-tunable transistor local oscillator.

7.5 MEASUREMENTS

The modern method of measuring basic properties for microwave transistors is by S (or scattering) parameters. This technique overcomes various problems associated with older methods of characterization by h- or y-parameters. The earlier difficulties were associated with the inaccessibility of the terminals of a transistor chip and the possibility of tuning or bias circuits presenting reactive terminations leading to oscillations at low frequencies. Further, open- or short-circuit terminations must be avoided if broadband swept frequency measurements are to be obtained. As errors become multiplied, calibration can be onerous, particularly in relation to the setting up of a consistent set of reference planes.

The small-signal response is defined in terms of the voltages and currents at the various terminals, and with the S-parameter representation, linear combinations of voltage and current are used as both independent and dependent variables. Thus, for a 2-port network,

$$\begin{bmatrix} b_1 \\ b_2 \end{bmatrix} = \begin{bmatrix} S_{11} & S_{12} \\ S_{21} & S_{22} \end{bmatrix} \begin{bmatrix} a_1 \\ a_2 \end{bmatrix} \quad (7.14)$$

where $a_1 = \dfrac{1}{2\sqrt{Z_0}}(v_1 + Z_0 i_1), \quad b_1 = \dfrac{1}{2\sqrt{Z_0}}(v_1 - Z_0 i_1)$

$a_2 = \dfrac{1}{2\sqrt{Z_0}}(v_2 + Z_0 i_2), \quad b_2 = \dfrac{1}{2\sqrt{Z_0}}(v_2 - Z_0 i_2)$

S_{11} and S_{22} are the input and output reflection coefficients; S_{21}^2 and S_{12}^2 are the forward and reverse insertion gains; $\angle S_{21}$ and $\angle S_{12}$ are the insertion phase shifts.

Lange [16] has discussed the application of this system to transistors as 2-port devices, operated with the common lead connected to earth. Unfortunately, the inductance introduced by this connection has important parasitic effects, particularly if the spacings are large, as is required in a 50 Ω system. A system of 3-terminal scattering parameters is to be preferred, as this considerably reduces the errors due to the common lead inductance. This system is described in a paper by Bodway [17], in which a design procedure is also given which, it is claimed, is particularly easy to apply.

In addition to S-parameters, it is usual to measure also various gain and stability factors. Details of these measurements may be found in the paper by Lange [16].

REFERENCES

1 BARDEEN, J., and BRATTAIN, W. H., "The transistor, a semiconductor triode", *Phys. Rev.*, **74**, p. 230 (1948).

2 TENZER, M., "Trends in semiconductor devices for microwave applications", *Electro-Tech.*, No. 10, p. 82 (1961).

3 LATHROP, J. W., "Semiconductor network technology", *Proc. Inst. Radio Engrs*, **52**, p. 1430 (1964).

4 PRITCHARD, R. L., "High-frequency power gain of junction transistors", *ibid.*, **43**, p. 1075 (1955).

5 EARLY, J. M., "Structure-determined gain-band product of junction triode transistors", *ibid.*, **46**, p. 1924 (1958).

6 KIRK, C. T., Jr., "A theory of transistor cut-off frequency (f_T) falloff at high current densities", *Trans. Inst. Radio Engrs*, **ED9**, p. 164 (1962).

7 JOHNSON, E. O., "Physical limitations on frequency and power parameters of transistors". *RCA Rev.*, **26**, p. 163 (1965).

8 DAS, M. B., *Microwave Potentialities of M.O.S.T., S.E.T. and Bipolar Transistors Compared*, (ASM memorandum, unpublished).

9 COOK, H. F., "Advances in microwave transistors and mixer diodes", *Proc. IERE/IEE Joint Symposium on Microwave Applications of Semiconductors*, Paper 18 (1965).

10 NEILSON, E. G., "Behaviour of noise figure in junction transistors", *Proc. Inst. Radio Engrs*, **45**, p. 957 (1959).
11 COOK, H. F., "Transistor noise figure", *Solid State Design*, **4**, No. 2 (1963).
12 TATUM, J. G., "Microwave transistor parameter trade-offs in circuit design", **6**, No. 9, p. 26; No. 10, p. 46; No. 11, p. 44 (1967).
13 COULTON, M., SOBEL, H., and ERNST, R. L., "Generation of microwave power by parametric frequency multiplication in a single transistor", *RCA Rev.*, **26**, p. 286 (1965).
14 GELNOVATCH, V., "Design of distributed transistor amplifiers at microwave frequencies", *Microwave J.*, **10**, No. 2, p. 41 (1967).
15 HERBERT, C. N., and CHERNEGA, J., "Broadband varactor tuning of transistor oscillators", *Microwaves*, **6**, No. 2, p. 28 (1967).
16 LANGE, J., "Modern tests for modern transistors", *ibid.*, **6**, No. 12, p. 38 (1967).
17 BODWAY, G. E., "Circuit design and characterization of transistors by means of three-port scattering parameters", *Microwave J.*, **11**, No. 5, p. 55 (1968).
18 LEE, M. C., *Microwave J.*, **12**, No. 2, p. 51 (1969).
19 VINCENT, B. T., *Digest of IEE G-MTT Conference*, p. 83 (1965)

8

Gunn-effect Devices

8.1 INTRODUCTION

Gunn-effect devices (or Gunn diodes) are named after J. B. Gunn, who, in 1963, whilst studying the properties of thin specimens of gallium arsenide and indium phosphide under high electric stress, discovered periodic fluctuations in the current passed by both materials when the applied voltage exceeded a certain critical value [1]. The particularly interesting feature of these experiments was that the frequency of oscillation could be made to lie in the microwave range by using specimens a few thousandths of an inch in thickness, and peak power outputs of several watts could be obtained with inputs of only a few volts. The mechanism responsible for this effect has since been shown to be that predicted earlier independently by Ridley and Watkins [2] and by Hilsum [3]. It is essentially that of the transfer of electrons from the lowest valley in the conduction band to valleys of higher energy where the mobility is less, leading to negative differential resistivity and travelling domains of high electric field within the semiconductor. The material of prime interest for such devices is gallium arsenide.

The main proof that the mechanism is the one indicated lies in the observation that the critical field is diminished as the energy separation between the valleys is reduced by applying pressure [4] or by alloying with gallium phopshide [5]. Gunn demonstrated the existence of travelling domains using an elegant capacitive probe technique [6], and their occurrence has also been confirmed indirectly from the observation of recombination radiation due to holes produced by avalanching [7]. Under certain conditions the domains may be immobilized, in which case a Gunn-effect device may be used as an amplifier [8, 9].

The electron-transfer mechanism requires that conduction states shall exist with energy levels a few times the thermal energy and in which the

Gunn-effect Devices

electron has high effective mass. Drifting electrons may acquire enough energy from an applied electric field to reach these states and, as they do so, suddenly increase their effective mass, with a consequent reduction in drift velocity. When the drift field exceeds the critical value electrons will undergo this process, and initially an accumulation layer of negative charge is formed at some nucleation centre near the cathode end of the specimen. This is quickly followed by the formation of a layer of positive charge and the dipole layer thus produced drifts through the semiconductor material as a unit structure with the saturation velocity of electrons in the material.

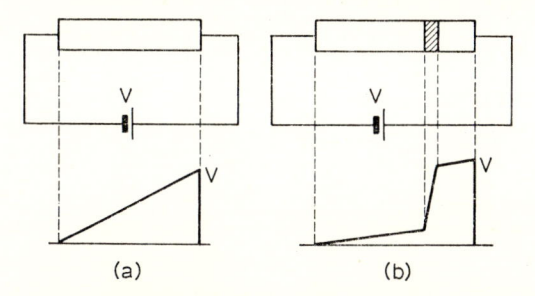

FIG 8.1 Potential distribution with and without domain
(a) Uniform field
(b) With domain

This constitutes a travelling domain, which becomes extinguished on reaching the anode, when the whole process can repeat itself. The effect on voltage distribution of increasing a direct voltage across the semiconductor specimen from one below to one above the value corresponding to the critcal field is illustrated in Fig. 8.1. Since the formation of a domain reduces the field outside it, only one domain forms at a time, although it may not always nucleate at the same point.

The current through the specimen initially rises in proportion to the applied voltage until the critical threshold value V_T is reached at which the current is I_T. If the voltage is made slightly greater than V_T the current rapidly drops to a value I_V, at which it remains for a time before rising, somewhat more slowly, to its original value I_T, the delay between current peaks corresponding approximately to the transit time of a domain through the specimen. The current waveform under such conditions is thus of the form shown in Fig. 8.2, although there may in practice be some deviation from this pattern. In general, the values of I_T and I_V are independent of the amount by which the applied voltage exceeds the threshold value, as is also the interval between current pulses. By suitable coupling, the specimen may be arranged to maintain oscillations in a resonant cavity tuned to a frequency such that one domain per cycle passes through the specimen.

The saturation velocity of electrons in gallium arsenide is about 10^7 cm/s, so that for the "transit-time" mode an effective specimen length of 10^{-3} cm would correspond to 10^{10} domains passing per second through the material, i.e. to a frequency of 10 GHz. This assumes that both the dielectric relaxation time and the transition time of electrons between high and low mobility states are short compared with the transit time of the domains.

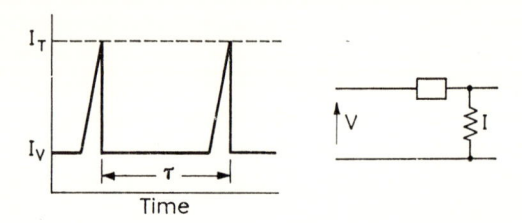

Fig 8.2 Current waveform for domain transit mode

The second of the above assumptions would certainly be valid for an oscillator operating at 10 GHz, since transitions between the states take place in a time of the order of 10^{-12} s. However, the dielectric relaxation time, given by the product of permittivity and resistivity, is about 10^{-11} s for 10 Ω-cm gallium arsenide at low fields. Further, the negative mobility associated with the high field domain is reported to be about 25 times smaller than the low field value, so that for high-resistivity material at microwave frequencies relaxation effects can significantly alter the build-up time and may even inhibit completely domain formation under certain conditions.

The critical field for gallium arsenide is about 3·6 kV/cm for short specimens, falling somewhat as the specimen length is increased. The minimum applied voltage for domain formation in a specimen 10^{-3} cm long would thus be 3·6 V, which may be compared with typical recommended operating voltages of around 6 V for 10 GHz Gunn-effect oscillators. The mode of operation first observed by Gunn, in which the period of oscillation is directly related to the specimen length, is called the "transit-time" mode, but several other modes have since been shown to be possible.

Gunn-effect Devices

Perhaps the most important basic advance since Gunn's original observations was the discovery at the Bell Telephone Laboratories of a mode of operation called LSA (limited space-charge accumulation) which in general demands a large component of steady bias and a large amplitude of superimposed voltage oscillation [10]. In this mode the time available for space-charge accumulation in each cycle is so short that domains cannot form and the device then operates as a simple negative resistance which may be connected into a cavity which is resonant at any chosen frequency, not dependent on the specimen length.

By using thicker material the power capability is much enhanced, limited only by thermal and voltage breakdown considerations. The LSA mode suggests the possibility of a high-power pulsed device which would rival the magnetron.

Gunn-effect oscillators have advantages in size, weight, simplicity of power supplies and in the cost of manufacture. They are reported to be less noisy than avalanche oscillators and have efficiencies of several per cent—much higher than for klystrons. The lifetime of devices under operating conditions has been a problem but is now largely overcome. Where high stability of frequency is required Gunn-effect oscillators are capable of injection-phase-locking, but wide tuning ranges are alternatively possible. For instance, in the transit-time mode a conventional 10 GHz oscillator can readily be tuned over 50 MHz, using a varactor diode, and in addition it has a mechanical tuning range of at least 1 GHz. A typical cavity for this band would have a volume less than 20 cm^3, and would weigh less than 30 g.

8.2 THEORY

The Gunn effect is dependent on there being a certain range in the electron-velocity/electric-field characteristic over which the differential mobility is negative, as indicated in Fig. 8.3 for *n*-type gallium arsenide. The origin of this negative resistance may be understood by reference to the conduction band structure for this material, illustrated in Fig. 8.4. The central valley at wave vector $k = 0$ contains electrons of effective mass ratio 0·067 and mobility of about 7500 cm^2/Vs. There are also satellite valleys, probably associated with zone boundaries, in which the effective mass ratio is 0·35 and the mobility around 150 cm^2/Vs, the minimum of these valleys being at 0·36 eV above the minimum of the central valley.

The first effect of applying an increasing electric field across a specimen of gallium arsenide is to increase the energy and momentum of electrons

Microwave Semiconductor Devices

in the central valley. When the field is high enough for electrons to gain an energy of 0·36eV they may remain in the central valley, but most of them will transfer to a satellite valley in which the density of states is high, the associated momentum change being produced by longtitudinal optical phonons. As a result, the electron mobility falls by a factor of about 50

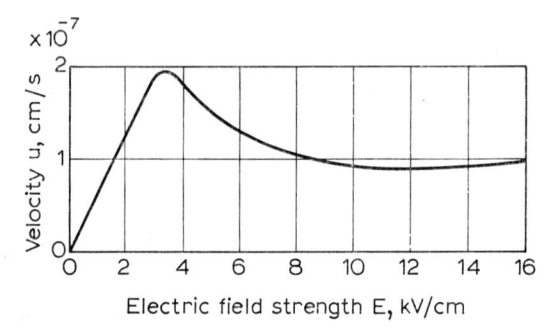

FIG 8.3 Static velocity v. field strength for GaAs

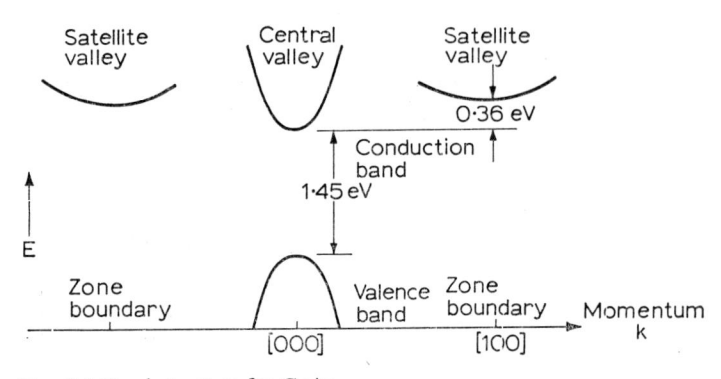

FIG 8.4 Band structure for GaAs

times, and this accounts for the region of negative differential mobility indicated in Fig. 8.3.

Scattering mechanisms modify the precise shape of the velocity/field characteristic, impurity and polar mode scattering being dominant at low fields but diminishing in importance at high values, thus facilitating the desired rapid increase in electron velocity with field when the latter reaches the critical value. The electron distribution is then effectively stabilized

Gunn-effect Devices

by the satellite valleys, in which electron mobility is largely governed at low fields by polar optical scattering and at high fields by deformation potential scattering, when the velocity becomes almost independent of the field.

The required properties of materials for Gunn-effect devices are first that a satellite valley should exist at an energy level difference W above a high-mobility central valley. The value of W must be large enough to prevent the thermal excitation of carriers from the valence to the conduction band but small enough to prevent field excitation between these bands occurring before the electron transfer mechanism becomes fully effective. In addition, a decrease in the scattering mechanisms should occur in the vicinity of the critical field to give a rapid transfer of electrons with a small increment of field and hence a steep negative slope. Compound semiconductors that have been shown to meet these requirements, in addition to those already referred to, include such materials as CdTe, ZnSe, $GaAs_xP_{1-x}$ and InAs (under pressure).

It is convenient to use the product nl (n = electron concentration, l = sample length) to distinguish between transferred-electron devices in which effects due to the contacts can be neglected and those in which they are dominant. The former is the situation for "long" samples in which $nl \gg 10^{12} \text{cm}^{-2}$, and the latter applies to "short" samples in which $nl \approx 10^{12} \text{cm}^{-2}$.

Domains originate at sites within the bulk semiconductor at which there are small variations in the space charge. Such fluctuations will normally decay exponentially with a time-constant which is the dielectric relaxation time. Since this is a negative quantity in a region in which the differential mobility is negative, a space charge fluctuation will grow in such a region until saturation occurs.

A stable field distribution will result, which can then travel through a "long" specimen unchanged in shape. Stable domains are of rounded triangular shape, with maximum fields in gallium arsenide up to 150 kV/cm, the field outside the domain being substantially uniform at some value less than the critical value. The Poisson and current-continuity equations must, of course, be satisfied throughout.

When the average field is above the critical value any further increase causes the peak field E_D within the domain to increase whilst the field E_R outside the domain decreases. Since the conduction current is controlled by E_R the differential resistance across the terminals of the device is negative in the presence of a domain. Butcher [11] has given a method for

Microwave Semiconductor Devices

obtaining the dynamic characteristic of E_D against velocity from the static characteristic. This is illustrated in Fig. 8.5, where the areas a and b are made equal.

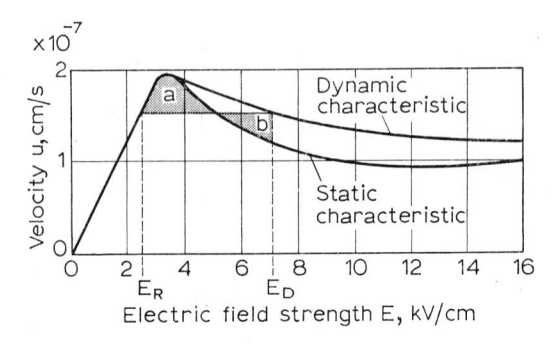

FIG 8.5 "Equal-areas" construction
(From Ref. 29)

Butcher derived an equal-areas rule, applying appropriate boundary conditions, from Poisson's equation,

$$\frac{\partial E}{\partial x} = \frac{e(n - n_0)}{\epsilon_0} \tag{8.1}$$

and the equation of conservation of total current density,

$$\frac{\partial J}{\partial x} = 0 \tag{8.2}$$

where $n = n_0$ at $E = E_R$; ϵ_r is the relative permittivity; and

$$J = neu - q\frac{\partial}{\partial x} Dn + \epsilon_r \frac{\partial E}{\partial t} \tag{8.3}$$

The three terms in this equation are the current contributions from conduction, diffusion and displacement. The argument proceeds by finding a differential equation for n as a function of E, to which the formal solution is obtained

$$\frac{n}{n_0} - \log_e \frac{n}{n_0} - 1 = \frac{\epsilon_r}{en_0 D} \int_{E_R}^{E} \left[(u - u_D) - \frac{n_0}{n}(u_R - u_D) \right] dE \tag{8.4}$$

It may further be shown that acceptable solutions require that $u_D = u_R$, i.e. the domain velocity and the electron drift velocity outside the domain

Gunn-effect Devices

are equal. The peak field within the domain is of such a magnitude that the integral in eqn. (8.4) vanishes when $u_R = u_D$, which implies that the two shaded areas in Fig. 8.5 should be equal.

It is possible to derive the domain voltage from the domain shape and the value of E_R obtained by the equal-areas construction. Knowledge of the domain voltage, together with the applied voltage and sample length, enables the domain parameters to be calculated.

If an external circuit is made to react on a Gunn-effect device in such a way that, when a domain reaches the anode the total field across the specimen is below threshold, then the renucleation of a domain will be delayed.

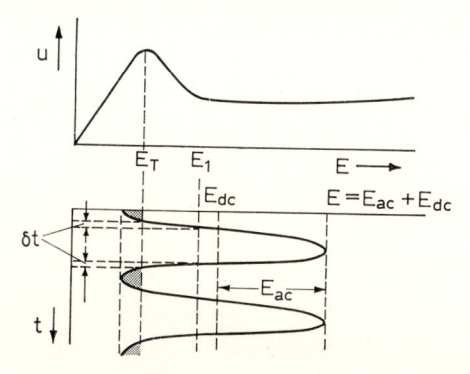

FIG 8.6 Conditions for LSA mode

Also, if the alternating field created within the specimen by an external circuit is large enough to drive the net field below a minimum sustaining value whilst a domain is in transit, then the domain will be extinguished.

Both of these effects can be utilized to permit the frequency of oscillation to be controlled by the circuit rather than by the length of the specimen over a range from about half to twice the transit frequency.

The time t_F taken for a domain to form is a function both of the dielectric relaxation time τ_D and of the sample length, since the domain capacitance has to be charged through the resistance of the sample. It may be shown [12] that $t_F \approx \sqrt{(\tau_D T)}$, where $T = 1/u_D$. If the transit time is less than t_F, domains are unable to form and the LSA mode may arise, the conditions for which are illustrated in Fig. 8.6. The time available for space-charge growth in each cycle δt must be small compared with the average negative dielectric relaxation time between E_T and E_1, and also the time spent below threshold at each cycle (shaded areas) must be long

Microwave Semiconductor Devices

enough to allow for dissipation of the space-charge growth which occurs as the field is swinging through the region of negative differential mobility. This implies that there is no net space-charge growth from cycle to cycle.

The doping/frequency ratio n/f, is a useful criterion for determining whether or not a device can be operated in the LSA mode. For large values of n/f it is not possible to meet the condition that $\delta t \ll \tau_D$, but if this ratio is too small the device operation cannot be efficient. Copeland [13] has suggested that a useful range for LSA operation is defined by

$$2 \times 10^5 > n/f > 10^4 \, \text{s/cm}^3$$

8.2.1 Power Transfer to an Oscillating Field*

The power transferred by the diode to a resonant cavity may be assessed through consideration of the current and voltage waveforms appearing across it. Provided the cavity Q-factor is high it is reasonable to suppose

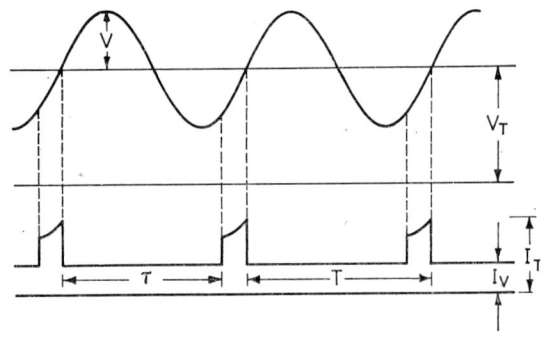

FIG 8.7 Modified current with cavity interaction
(*From Ref.* 14)

that the alternating voltage across the diode will be purely sinusoidal. Near to resonance, the cavity, together with the diode lead inductance and capacitance, may be represented simply as a susceptance in parallel with a conductance.

Consider a specimen which is biased just to the threshold voltage. After a short time the total voltage appearing across it will consist of the bias voltage together with an alternating component at the resonant frequency. Nucleation of a domain will occur as the alternating component passes through zero and is positive going, and the current will fall from its

* Sections 8.2.1 and 8.2.2 are based on a treatment by Robson and Mahrous [14].

threshold value I_T to the valley level I_V at which it will remain until the domain reaches the anode. At this time it will revert to the value I_T provided that the instantaneous voltage is at least equal to the threshold value. However, should the period of oscillation of the cavity be greater than the domain transit time τ, the situation is altered and the specimen will remain ohmic until the total applied voltage again exceeds the threshold value. Figure 8.7 shows the current waveform corresponding to these conditions. Fourier analysis of this waveform gives the fundamental in-phase component as

$$I_P = -\frac{2}{\pi}(I_T - I_V)\sin^2\frac{\omega\tau}{2} + \frac{V}{R}\left[\frac{1-\omega\tau}{2\pi} + \frac{\sin 2\omega\tau}{4\pi}\right] \qquad (8.5)$$

and the quadrature component as

$$I_q = -\frac{1}{\pi}(I_T - I_V)\sin\omega\tau - \frac{V\sin^2\omega\tau}{2\pi R} \qquad (8.6)$$

where R is the low-field resistance of the specimen.

It is assumed above for simplicity that the rise and fall of the current pulse is very short compared with a period T, implying a situation in which the thickness of a domain is much less than the sample length, which will be true if the resistivity is not too high. An essential condition for this mode of operation to be tunable by the cavity is that

$$\frac{T}{2} < \tau < T \qquad (8.7)$$

If this condition is not satisfied a domain will be retriggered immediately after reaching the anode and the arrangement will not be tunable. If, during the transit of a domain, the voltage falls below the value $I_V R$ the domain is extinguished and from that point onwards ohmic operation ensues until the voltage again exceeds the threshold value.

The domain cannot be extinguished in transit if

$$\frac{V}{R} < I_T - I_V \qquad (8.8)$$

From the above inequality together with eqn. (8.5) it is easy to show that for $T/2 < \tau < T$ the fundamental in-phase current I_P is always negative and oscillations are sustainable. The maximum tuning range predicted

173

Microwave Semiconductor Devices

for this particular mode of operation is therefore one octave, from $1/2\tau$ to $1/\tau$. Other tunable modes are possible which satisfy the inequality

$$(n - \tfrac{1}{2})T < \tau < nT \tag{8.9}$$

in which n is a positive integer. Such modes have frequencies higher than the reciprocal of the transit time but are associated with poor efficiency.

If the inequality (8.8) is not satisfied it may be shown that a domain can be extinguished over a very small range of τ, but this is not of practical importance.

From eqn. (8.5), noting that $V = -I_P R_L$, where R_L is the cavity load resistance, a steady-state relationship may be obtained between V, $\omega\tau$, R and R_L, namely

$$\frac{V}{R(I_T - I_V)} = \frac{\dfrac{2}{\pi}\sin^2\dfrac{\omega\tau}{2}}{\left(1 + \dfrac{R}{R_L}\right) - \dfrac{\omega\tau}{2\pi} + \dfrac{\sin 2\omega\tau}{4\pi}} \tag{8.10}$$

Dependent on whether or not the two sides in eqn. (8.10) are greater or less than unity, it may or may not be possible to extinguish a domain in

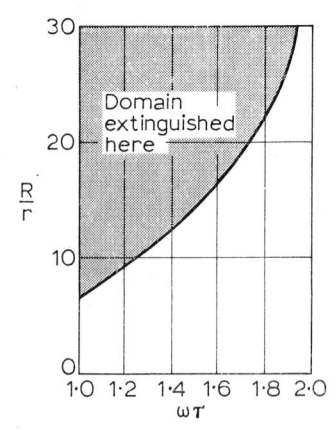

FIG 8.8 Load resistance to extinguish domain
(From Ref. 14)

transit. The range of load resistance over which a domain can be extinguished has been plotted by Robson and Mahrous [14] and their diagram is reproduced in Fig. 8.8.

Gunn-effect Devices

8.2.2 Efficiency

As a conservative estimate we may take the input power to be $I_T I_V R$.

The amplitude of the alternating voltage developed across the specimen will never much exceed $(I_T - I_V)R$. Substituting this value into eqn. (8.5) and noting that the maximum fundamental component of current occurs at $\omega\tau = \pi$, the maximum efficiency is found to be

$$\eta_{max} = 6 \cdot 8(1 - I_V/I_T)(I_T/I_V - 1) \text{ per cent} \qquad (8.11)$$

In practice I_T/I_V is believed to be about 2 for gallium arsenide, which suggests a maximum efficiency of 3·4%. This fairly low value results from a

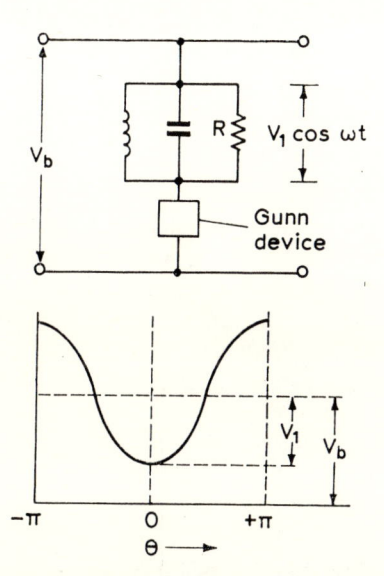

FIG 8.9 Conditions relevant to LSA mode

simplifying assumption that the bias field is equal to the threshold field. F. L. Warner [15] has extended the calculations to arbitrarity large bias fields and obtained $\eta_{max} = 7 \cdot 2\%$.

For the LSA mode of operation the efficiency may be gauged with reference to Fig. 8.9, in which the range of primary interest centres on the negative extreme of the circuit voltage excursion, which is taken as determining the zero for the measurement of phase angle.

The power output is given by

$$P_0 = \frac{1}{\pi} \int_0^\pi IV_1 \cos\theta\, d\theta = \frac{\phi}{\pi} V_b I_0 \tag{8.12}$$

Now, ϕ is given by

$$\phi = \frac{V_1}{V_p} \int_0^\pi y \cos\theta\, d\theta \tag{8.13}$$

where $y = (u/u_0) - 1$. The ratio v/v_0 may be obtained from the velocity/field characteristic as derived theoretically by Butcher and Fawcett [16]. The velocity u is approximately equal to u_0 for a field E_0 of about $1 \cdot 1\,\text{kV/cm}$, u_0 being the saturation electron velocity of about $8 \times 10^6\,\text{cm/s}$.

Using the above values, it may be shown that

$$\phi = \frac{3 \cdot 5 \sqrt{(E_b - E_0)}}{E_b} \tag{8.14}$$

For high values of V_b the power input is approximately $V_b I_0$, so that the efficiency becomes

$$\eta = \frac{\phi}{\pi + \phi} \tag{8.15}$$

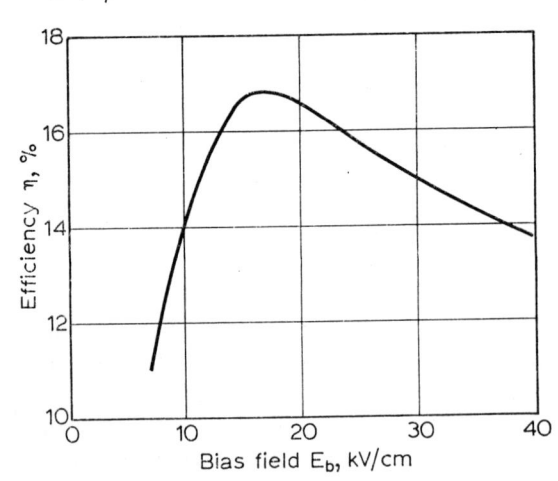

FIG 8.10 Efficiency for LSA mode v. bias field
(J.E. Curran—private communication)

Gunn-effect Devices

This is plotted against E_b in Fig. 8.10 for $E_{min} = E_0$ and is seen to rise to a maximum of 17%, which may be compared with about 9% achieved in a practical device.

8.2.3 Dimensional Requirements

For the LSA mode of operation there is a limitation of $10^9/f$ centimetres on one of the dimensions perpendicular to the electric field (the other perpendicular is not subject to this limitation). This is a result derived by Copeland [17, 18], who also quoted experimental evidence of such a limitation. There is no limitation on length as measured in the direction of electric field, other than that of being reasonably small compared to the wavelength.

8.3 STATE OF THE ART

Table 8.1 on the state of the art with Gunn-effect oscillators was prepared in February, 1970.

TABLE 8.1

Frequency (GHz)	Power Output (W)	Efficiency (%)	Mode of Operation	Reference
1·75	6000·0	14·6	LSA pulsed	19
3·05	200·0	29·0	Gunn pulsed	20
7·0	2100·0	4·0	LSA pulsed	19
7·7	350·0	3·0	LSA pulsed	21
8·3	19·0	22·0	Gunn pulsed	22
8·7	0·78	2·5	Gunn c.w.	23
13·3	0·17	3·0	Gunn c.w.	24
15·0	10·0	10·0	Gunn pulsed	24
24·8	0·12	5·2	Gunn c.w.	25
30·0	20·0	7·5	LSA pulsed	26
44–51	0·020	0·7	LSA c.w.	27
50·0	3·5	0·5	LSA pulsed	28
84–88	0·020	2·0	LSA c.w.	27

8.3.1 Noise Properties

Both a.m. and f.m. noise properties of Gunn oscillators are significantly better than for avalanche oscillators, certainly at frequencies in excess of

100 kHz away from the carrier. The former may, in fact, be used satisfactorily in single-ended mixers as local oscillators, with intermediate frequencies of around 30 MHz.

Although amplification with Gunn-effect devices can be achieved, both with stable and oscillating specimens, noise figures so far quoted are very poor (about 20 dB) [30, 31, 32].

8.4 PREPARATION OF GALLIUM ARSENIDE MATERIALS

The following description of gallium arsenide processing methods is taken with permission from an article by J. M. Woodall and K. L. Lawley, which appeared in *Electronics*, 13th November, 1967 (copyright McGraw-Hill Inc. 1967).

The development of techniques for the preparation of suitable gallium arsenide material accounts for much of the effort that has gone into the development of Gunn-effect devices, a major obstacle being the relative impurity of the material compared with silicon and germanium. Epitaxial material has been shown to give advantages over bulk-grown material, but even here the quality of the device is very dependent on the quality of the bulk-grown substrates.

We shall therefore first discuss methods of crystal growth. The two methods principally employed are those of Czochralski and of Bridgman.

Two variants of the Czochralski technique are currently in use. In the first a sealed system is used to contain the arsenic vapour, temperatures in excess of 614°C being employed to maintain the proper pressure of arsenic. Since the arsenic vapour is highly reactive it is usual to lift the seed by means of iron parts that are encapsulated in quartz and coupled magnetically to some external lifting arrangement. Alternatively it is possible to use non-reactive high-temperature seals and feed-through for the lifting arrangement. It is usual to employ resistance heating to maintain the arsenic vapour pressure together with r.f. heating for the melt. In the other version of Czochralski's method the melt is encapsulated in some vitreous non-reactive material such as boric oxide and an inert gas at a pressure of about 100 kN/m^2 is used to prevent arsenic escaping from the melt.

There are also two variants of the method due to Bridgman, horizontal and vertical, but the latter is not generally used for semiconductor materials. In the horizontal method the melt is caused to solidify progressively into a single crystal structure by passing it slowly through a zone of uniform temperature gradient which spans the melting point. Growth in a preferred crystal plane direction may be obtained by seeding. This method generally

Gunn-effect Devices

produces crystals of better uniformity and higher purity, the uniformity resulting from the fact that the motor-driving mechanism can be simpler operating at room temperature.

For the Czochralski method, typical high-purity results are about 4×10^{15} impurity atoms/cm^3 and electron mobility of 7000cm^2/Vs. Corresponding figures for the Bridgman method are 5×10^{14} atoms/cm^3 and 8000cm^2/Vs. With suitable precautions, both methods can give dislocation-free material.

The production of epitaxial gallium arsenide has been a key process in improving Gunn-device technology. Since relatively low substrate temperatures can be employed, pure crystals can be grown which have the maximum electron mobility, which for Gunn-effect material is about 10000cm^2/Vs at 300 K and as high as 130000cm^2/Vs at 77 K [33].

There is a lack of suitable compounds of gallium arsenide which are gaseous at normal temperature and pressure, which makes it a more difficult matter to grow films of gallium arsenide compared with silicon or germanium. The simplest practical method is to use bulk gallium arsenide as a source material and to transport this via either water vapour or a chloride. With water vapour, the source material is maintained at about 1050°C, the reaction products being hydrogen, arsenic vapour and the volatile suboxide of gallium. The substrate undergoing deposition is situated at a cooler part of the reactor (c. 1000°C) where the reaction is reversed and a single-crystal film is deposited.

Temperatures are lower with the chloride process, around 800°C for the source and 750°C for the substrate, but films produced in this way tend to be less pure than when water vapour is used, since impurities from the source and feed-gas are all too easily transferred to the film as chlorides.

An alternative technique, which gives somewhat purer films, is to use a source of pure gallium to give a gaseous gallium compound in conjunction with HCl or chlorine. Arsenic trichloride, or arsine, may be used to provide the arsenic. Deposition of the gallium arsenide substrate occurs at a temperature of 750–800°C. Doping materials for *n*-type layers include hydrogen sulphide and selenide. For *p*-type films, zinc metal or dimethyl zinc have been employed.

Another method, which probably gives the purest films, uses a source formed by the exposure of liquid gallium to arsenic trichloride. A skin of gallium arsenide forms on the gallium at the source operating temperature and deposition takes place from the reaction products at a substrate maintained at about 750°C.

Microwave Semiconductor Devices

Various precautions are found necessary to overcome snags which arise in practice. For instance, a p-type layer may sometimes occur between the substrate and film, which can be avoided if the substrate is first given a prolonged soak in hot potassium cyanide. The value of this process is believed to be due to the removal of copper from the substrate surface. Also, maintenance of a temperature gradient in the substrate zone is found to help in obtaining layers of reproducible thickness. Homogeneity is important and field-dependent trapping centres are undesirable.

8.5 DIODE STRUCTURE

The type of structure which is most widely used for Gunn oscillators is known as "sandwich" or "longitudinal" and is illustrated in Fig. 8.11.

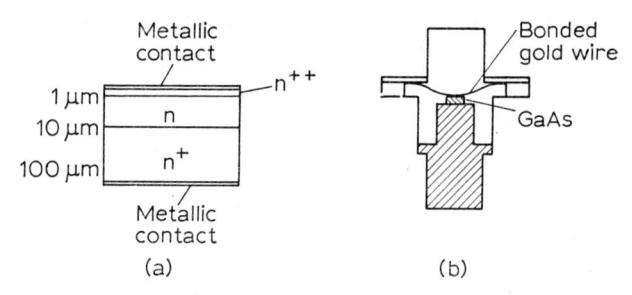

FIG 8.11 Longitudinal structure and typical "pill" mounting arrangement

Basically, a thin layer of epitaxially deposited high-purity gallium arsenide is used in conjunction with a high-conductivity substrate. Contact to the high-purity layer may be made through a still thinner layer of very heavily doped material, as shown, but this layer is often omitted and the metallized contact material is then usually an evaporated and fired-in silver-tin alloy. This acts as the cathode and the substrate is made into the anode. More recently, planar structures have been developed, aimed at devices more suitable for mass production. With these planar (or "transvere") devices, the epitaxial layer is deposited on a substrate of semi-insulating gallium arsenide. Surface passivation with silicon dioxide is difficult, but the nitride process has given successful results. For contacts, alloys are the only materials which satisfy the requirements demanded by processing and performance; their composition and alloying cycle is generally critical. Besides silver-tin, other materials such as Au-Ge, In, In-Ag, etc., are being used for the contacts.

Gunn-effect Devices

8.6 MOUNTING ARRANGEMENTS

An example of a pill-encapsulated Gunn oscillator used directly across waveguide appropriate to a wavelength of 1 cm is illustrated in Fig. 8.12(*a*). At lower frequencies a coaxial arrangement may be employed, as outlined at (*b*).

A voltage-tunable Gunn oscillator making use of a device cross-section increasing from the cathode was first reported by M. Shoji [34]. Subsequently Jeppsson *et al.* [35] described a special configuration of such an

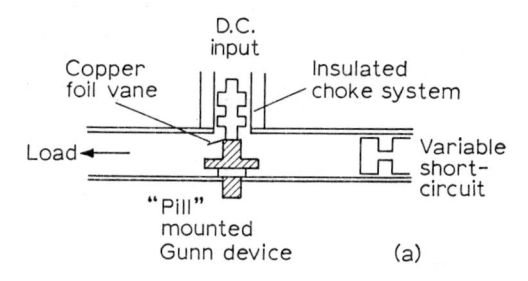

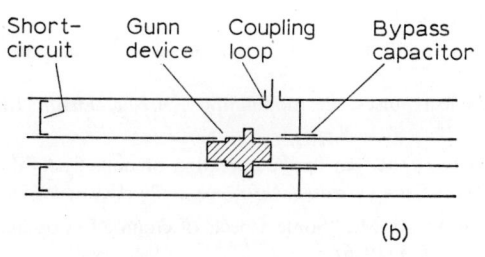

FIG 8.12 Oscillator circuit arrangements
    (*a*) Waveguide (J. Bott—*private communication*)
    (*b*) Coaxial line (*From Ref.* 14)

oscillator having planar concentric contacts. As a domain expands on its way to the anode the current increases, with the result that the voltage across the domain falls, the domain collapsing as the voltage goes below threshold. A voltage tuning range exceeding an octave has been claimed in the frequency range 1·5–8·5 GHz.

REFERENCES

  1 GUNN, J. B., "Microwave oscillations of current in III–V semiconductors", *Solid State Commun.*, **1**, p. 88 (1963).

181

2 RIDLEY, B. K., and WATKINS, T. B., "The possibility of negative resistance in semiconductors", *Proc. Phys. Soc.*, **78**, p. 293 (1961).

3 HILSUM, C., "Transferred electron amplifiers and oscillators," *Proc. Inst. Radio Engrs*, **50**, p. 185 (1962).

4 HUTSON, A. R., JAYARAMAN, A., CHYNOWETH, A. G., CORIELL, A. S., and FELDMAN, W. L., "Mechanism of the Gunn effect from a pressure experiment", *Phys. Rev. Letters*, **14**, p. 639 (1965).

5 ALLEN, J. W., SHYAM, M., CHEN, Y. S., and PEARSON, G. L., "Microwave oscillations in $GaAs_xP_{1-x}$ alloys", *Appl. Phys. Letters*, **7**, p. 78 (1965).

6 GUNN, J. B., "Instabilities of current and of potential distribution in GaAs and InP", *Proc. of Symposium on Plasma Effects in Solids* (Academic Press, 1964).

7 CHYNOWETH, A. G., "Current understanding of h.f. instabilities in bulk semiconductors", *Digest of International Solid State Circuits Conference*, p. 80 (IEEE, 1966).

8 THIM, H. W., BARBER, M. R., HAKKI, B. W., KNIGHT, S., and UENOHARA, M., "Microwave amplification in a dc-biased bulk semiconductor", *Appl. Phys. Letters*, **7**, p. 167 (1965).

9 KROEMER, H., "External negative conductance of a semiconductor with a negative differential mobility", *Proc. IEEE*, **53**, p. 1246 (1965).

10 COPELAND, J. A., "A new mode of operation for bulk negative resistance oscillators", *Proc. IEEE*, **54**, p. 1479 (1966).

11 BUTCHER, P. N., "The Gunn effect", *Reports on Progress in Physics*, **30**, Pt. 1, p. 97 (1967).

12 HARTNAGEL, H., "Gun-effect pulse-code modulation", *Archiv. Elektr. Ubertragung*, **22**, p. 225 (1968).

13 COPELAND, J. A., and SPIWAK, R. R., "LSA operation of bulk GaAs diodes", *Digest of International Solid State Circuits Conference*, p. 26 (IEEE, 1967).

14 ROBSON, P. N., and MAHROUS, S. M., "Some aspects of Gunn-effect oscillators", *Radio Electron. Engr*, **30**, p. 345 (1965).

15 WARNER, F. L., "Extension of the Gunn-effect theory given by Robson and Mahrous", *Electron. Letters*, **2**, p. 260 (1966).

16 BUTCHER, P. N., and FAWCETT, W., "Calculation of the velocity/field characteristics for gallium arsenide", *Phys. Letters*, **21**, p. 489 (1966).

17 COPELAND, J. A., "LSA oscillator diode theory", *J. Appl. Phys.*, **38**, p. 3096 (1967).

18 COPELAND, J. A., "Doping uniformity and geometry of LSA oscillator diodes", *Trans. IEEE*, **ED-14**, p. 497 (1967).

19 JEPPSSON, B., and JEPPESEN, P., "A high-power L.S.A. relaxation oscillator", *Proc. IEEE*, **57**, p. 1218 (1969).

20 REYNOLDS, J. F., BERSON, B. E., and ENTROM, R. E., "High-efficiency transferred-electron oscillators", *ibid.*, p. 1692.

21 KENNEDY, W. K., EASTMAN, L. F., and GILBERT, R. J., "L.S.A. operation of large-volume bulk GaAs samples", *Trans. IEEE*, **ED-14**, p. 500 (1967).

22 CALIFANO, F. P., "High-efficiency X-band oscillators", *Proc. IEEE*, **57,** p. 251 (1969).
23 NARAYAN, S. Y., ENSTROM, R. E., and GOBAT, A. R., "High-power c.w. transferred-electron oscillators", *Electron. Letters*, **6,** p. 17 (1970).
24 EDRIDGE, A. L., MYERS, F. A., DAVIDSON, B., and BASS, J. C., "Gunn-effect oscillators". Paper presented at 1969 European Microwave Conference, London. (Proceedings to be published.)
25 FANK, F. B., and DAY, G. F., "High c.w. power K-band Gunn oscillators", *Proc. IEEE*, **57,** p. 339 (1969).
26 EASTMAN, L. F., "Generation of high-power microwave pulses using gallium arsenide". Paper presented at 1969 European Microwave Conference, London. (Proceedings to be published.)
27 COPELAND, J. A., "C.W. operation of L.S.A. oscillator diodes", *Bell Syst. Tech. J.*, **46,** p. 284 (1967).
28 CAMP, W. O., and KENNEDY, W. K., "Pulse millimeter power using the L.S.A. mode", *Proc. IEEE*, **56,** p. 1105 (1968).
29 BUTCHER, P. N., *Phys. Letters*, **19,** p. 546 (1965).
30 THIM, H. W., and BARBER, M. R., "Microwave amplification in a GaAs bulk semiconductor", *Proc. IEEE*, **55,** p. 718 (1967).
31 THIM, H. W., and LEHNER, H. H., "Linear millimeter wave amplification with GaAs wafers", *ibid.*, p. 517.
32 THIM, H. W., "Linear microwave amplification with Gunn oscillators", *Trans. IEEE*, **ED-14,** p. 517 (1967).
33 REID, F. J., ROBINSON, L. B., KANG, C. S., GREENE, P. E., and SOLOMON, R., Papers presented at 1968 International GaAs Conf., Dallas, Texas. (Proceedings to be published.)
34 SHOJI, M., "A voltage-tunable Gunn-effect oscillator", *Proc. IEEE*, **55,** p. 130 (1967).
35 JEPPSSON, B., *et al.*, "Voltage tuning of concentric planar Gunn diodes", *Electron. Letters*, **3,** p. 498 (1967).

9

Avalanche Diode Oscillators

9.1 INTRODUCTION

Avalanche diode oscillators rely on the effect of voltage breakdown across a reverse-biased p–n junction to produce, over part of an applied voltage cycle, a supply of holes and electrons, one or other type of carrier then being subjected to a drift field at right angles to the junction and along the semiconductor specimen. The essential feature is that the field pattern and drift distance should together lead to an effective negative resistance permitting microwave oscillations or amplification with the diode mounted in an appropriate cavity.

The possibility of producing high-frequency oscillations from a reverse-biased p–n junction device operating under avalanche breakdown conditions was first discussed in 1958 by Read [1], who gave a detailed analysis relating to an n^+–p–i–p^+ structure. General interest in avalanche oscillators was not aroused, however, until February, 1965, when Johnston, De Loach and Cohen [2] reported that microwave oscillations had been obtained up to 24 GHz on a pulse basis from simple p^+–n structures. The first report of oscillations from a true Read structure came almost immediately afterwards, when Lee et al. [3] described harmonic mode oscillations up to 540 MHz. Later in 1965 Burrus [4] obtained oscillations up to 85 GHz from ordinary diffused-junction devices operated in an avalanche breakdown condition. The ability of an avalanching diode to provide amplification was quickly demonstrated, and as early as September, 1965, Napoli and Ikola [5] reported power gains of up to 10 dB at 8·75 GHz with a dynamic range of about 10 dB. The main interest in avalanche diodes, however, remains in their potential as sources of microwave power.

Efforts to improve avalanche oscillators have been mainly directed towards increasing the power output and improving the noise performance. A significant development occurred in 1967, when Prager, Chang and

Avalanche Diode Oscillators

Weisbrod [6] of R.C.A. Laboratories discovered an anomalous mode of operation which has led to high power at high efficiency at the lower microwave frequencies. The mode is anomalous in that oscillations can occur at small transit angles down to a few tenths of a radian and efficiencies are observed up to about twice that predicted by the model due to Read. A useful review of such diodes has been given by Chang [7]. The term TRAPATT has been applied to such devices (Trapped Plasma Resonance Mode). Compared with Gunn devices, the power output of avalanche diodes can be much higher, 4·7 W at 13·3 GHz having been obtained under c.w. conditions with a silicon avalanche oscillator mounted on a diamond heat sink [8]. Under pulsed conditions 30 W has been obtained at 8 GHz [9], but this is an order of magnitude less than the 350 W quoted in Chapter 8 for a Gunn diode operating in the pulsed LSA mode.

Both a.m. and f.m. noises tend to be high for avalanche oscillators. Satisfactory operation of receivers using such devices as local oscillators is nevertheless possible in conjunction with balanced mixers, which enables the a.m. noise to be adequately suppressed. Although f.m. noise is not altered by such arrangements, there are many systems for which this particular form of noise is unimportant. It has been claimed by Varian Associates that a significant improvement in a.m. noise can be achieved with a mesa structure, and that such a device will give virtually identical receiver performance as with a backward-wave oscillator used as the local oscillator. Improved noise performance can also be obtained by using a planar epitaxial structure in which the diffusion is carried through almost to the substrate [10]. As the junction approaches the substrate there is a reduction in the breakdown voltage and the avalanche process can be made to occur fairly uniformly over the depletion region, which is made to "punch through" to the substrate.

Strong electronic tuning effects are possible with avalanche structures through control of the avalanching direct current, tuning ranges of at least 10% being possible at 10 GHz [11]. Tuning by static pressure has also been demonstrated [12], but this effect is not of such general importance. Another demonstrated feature is the ability to operate two or more avalanche devices in parallel [13]. The product of power and (frequency)^2 is more than two orders of magnitude greater than can be obtained with transistors, and values of about 4000 W-GHz^2 have already been achieved, as may be seen from Table 9.1 (overleaf).

Avalanche diodes have been made with various impurity profiles from abrupt to graded, mesa and planar structures, while silicon, gallium arsenide

Microwave Semiconductor Devices

TABLE 9.1

Frequency (GHz)	Power output (W)	Efficiency (%)	Mode of operation	Diode material	Reference
0·425	435·0	22·0	pulsed	Si	7
0·450	450·0	43·0	c.w.	Ge	24
0·775	180·0	60·0	pulsed	Si	7
1·05	420·0	33·0	pulsed	Si	7
3·0	7·5	40·0	pulsed	Ge	21
6·0	0·62	12·1	c.w.	Ge	25
6·5	1·3	3·6	pulsed	GaAs	26
6·8	0·38	7·8	pulsed	GaAs	26
8·0	30·0	∼4·5	pulsed	Si	9
8·5	0·32	5·5	c.w.	Ge	27
9·6	0·96	12·5	c.w.	GaAs	28
13·3	4·7	∼8·0	c.w.	Si	8
13·8	2·7	10·2	c.w.	Si	29
16·5	2·1	5·0	pulsed	GaAs	30
21·0	8·8	4·5	pulsed	Si	9
27·0	0·4	2·0	pulsed	GaAs	31
50·0	0·64	10·0	c.w.	Si	32
107·0	0·74	3·2	c.w.	Si	32

and germanium have been used as the semiconductor materials. In due course it will doubtless emerge which one is the best for any given application. Although silicon appears to be superior in terms of power output, gallium arsenide and germanium are both superior in terms of noise. Noise performance comparable to that of a good single-cavity reflex klystron is said to be readily available with germanium, which is considerably easier to control than gallium arsenide.

9.2 THEORY OF THE READ DIODE

Read's original proposal was for an $n^+\text{-}p\text{-}i\text{-}p^+$ structure with a depletion layer extending to the p^+ back contact at breakdown. The structure and field distribution are illustrated in Fig. 9.1. The operation depends on a combination of avalanche breakdown and transit-time effects giving an effective phase difference between current and voltage across the structure of 180° and hence a negative resistance.

Avalanche Diode Oscillators

The n^+–p junction is reverse biased to a value well above that required for punch-through, so that a space-charge region is made to extend from the n^+–p junction to the i–p^+ junction. Maximum field occurs at the n^+–p junction, and when this is made to exceed the avalanche breakdown value (several hundred kilovolts per centimetre), internal secondary emission of hole-electron pairs occurs, the electrons immediately entering the n^+-region and the holes drifting to the right across the depletion region. Throughout

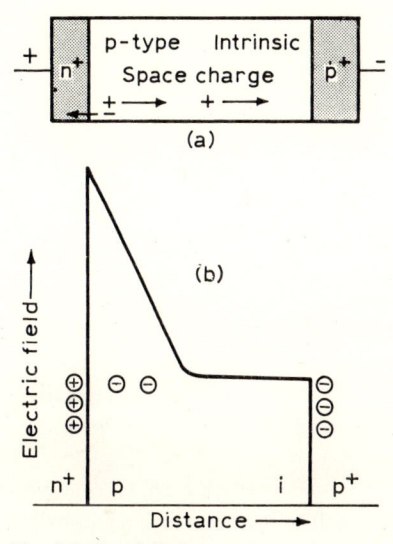

Fig 9.1 Read diode
(a) Structure
(b) Electric field under reverse bias
(From Ref. 1)

this region the field is of the order of 5 kV/cm, which is high enough to cause the carriers to move with a constant velocity independent of the field, about 10^7 cm/s for silicon. For a space-charge region of width W the transit time τ of holes moving across it is therefore given by $W \times 10^{-7}$ s.

A current $I_e(t)$ is produced in the external circuit by the holes moving across the space-charge region, where I_e is the average current within this region and equal to $1/\tau$ times the total charge of the drifting holes. If a pulse of holes of charge δQ is suddenly generated at the n^+–p junction, a constant current of magnitude $I_e = \delta Q/\tau$ will be made to flow in the external circuit. On average, the external current $I_e(t)$ will be delayed by $\tau/2$ relative to the current $I_0(t)$ which is generated at the n^+–p junction. It is shown

187

Microwave Semiconductor Devices

below that $I_0(t)$ is retarded by $\pi/2$ in relation to the alternating voltage, and therefore for a total delay of one half-cycle the transit-time delay $\tau/2$ must also correspond to a lag of $\pi/2$. The external circuit should therefore be tuned to a frequency of $1/2\tau$.

In addition to the conduction current $I_e(t)$ there is a displacement current I_c which supplies the variation in charge at the ends of the space-charge region at which the field changes abruptly. As this component of current leads the voltage by $\pi/2$ it contributes nothing to the power.

Avalanche multiplication is a sensitive non-linear function of the electric field, and the doping profile is aimed to give the field in the launching region a relatively sharp peak, confining the multiplication process to a narrow region at the n^+–p junction. The rate at which hole-electron pairs are generated is $\alpha v n$, where α is the ionization rate, and v and n represent velocity and electron concentration. As α is defined as the average number of pairs produced by each electron in moving through unit distance, the average distance between iozinizing collisions is $1/\alpha$. For the values of field considered by Read (less than 600 kV/cm), α is a function of the field and will be written as $\alpha(E)$. This relationship was originally studied by McKay [14], who found that to a first approximation $\alpha(E) \approx E^m$, where $m = 6$. Read, for simplicity, assumed equal ionization rates for electrons and holes, an assumption which is more nearly valid for germanium and gallium arsenide than for silicon.

At the breakdown field E_c every hole-electron pair on average produces one other pair, so that the current is self-sustaining. By biasing the diode so that the peak field is greater than E_c during the positive half of the voltage cycle and below it during the reverse half, the generated current $I_0(t)$ is built up during the positive half and decays in the reverse half, thereby lagging in phase by $\pi/2$. It is assumed here that the voltage and field vary in phase with each other, which is nearly true provided the current is not too large and space charge can be neglected.

The frequency of oscillation is determined by the capacitance of the diode and the inductance of its cavity. We have noted that $f = 1/2\tau$, and writing $\tau = W/v$, it follows that

$$f = \frac{v}{2W} \tag{9.1}$$

where W is the width of the space-charge region. Thus, for a frequency of 10 GHz, W is required to be about $5\,\mu\mathrm{m}$. Since the capacitance of such a

Avalanche Diode Oscillators

depletion layer in silicon is approximately $1{\cdot}06\,\text{pF/cm}^2$, the associated admittance would be about $60\,\text{S/cm}^2$. A possible cavity as originally proposed by Read is shown in Fig. 9.2.

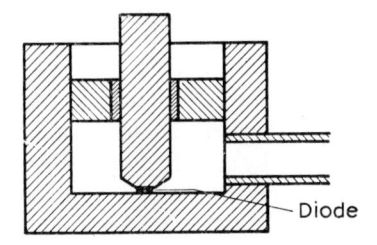

FIG 9.2 Oscillator cavity as proposed by Read
(*From Ref.* 1)

For a circular diode of radius a the impedance is given by $(\pi a^2 \omega C)^{-1}$, which must be equated to the impedance ωL of the cavity, leading to

$$\left(\frac{a}{W}\right)^2 = \frac{9{\cdot}5 \times 10^3}{\omega L} \text{ for silicon} \qquad (9.2)$$

Since ωL cannot readily be made less than about $10\,\Omega$, the radius is restricted to about $30W$, i.e. about $0{\cdot}15\,\text{mm}$ for a diode to operate at $10\,\text{GHz}$.

When the diode is oscillating in a cavity, although the field may vary by less than 20%, because of the highly non-linear avalanching process, the current will vary by orders of magnitude. The conduction current I_c will approach a square wave, being virtually zero during the positive voltage swing and substantially constant during the negative half of the cycle. The amplitude of variation of I_c will be approximately equal to the direct current I_d, since the latter is the average conduction current. The power output is proportional to the product of alternating voltage amplitude V_a and the bias current. If the latter is maintained constant, since the energy stored per cycle is proportional to V_a^2, the negative Q-factor of the diode is proportional to V_a. This condition gives a stable operating point, since any tendency for the amplitude to increase above a certain value is countered by the energy lost to the cavity increasing faster than the energy delivered by the diode, and vice versa.

The total power output for a diode of circular area πa^2 may readily be shown to be $2a^2 I_d V_a$. The maximum value of a is given by eqn. (9.2), and the maximum permissible value of I_d is limited by space-charge considerations, as indicated below.

Microwave Semiconductor Devices

The primary effect of the space charge associated with the moving carriers is to flatten the field distribution shown in Fig. 9.1 and thereby to reduce the generation of current carriers when the peak field is reduced below the sustaining value E_c. It follows that $I_0(t)$ will reach its peak value before the middle of the voltage cycle, thereby reducing the delay and hence the power. Thus beyond a certain value of bias current the power output will fall as the carrier space charge begins to spoil the phase relationship. The carrier space charge is given by τI_d whereas the charge producing the voltage variation is CV_n. Read showed that the current and voltage do not differ in phase appreciably from the desired 180° shift, provided that $\tau I_d \not> \frac{1}{2}CV_n$. Since the period of oscillation is 2τ, the maximum bias current which can profitably be used is given by

$$I_d = \frac{\omega}{2\pi} CV_a \tag{9.3}$$

Using eqns. (9.2) and (9.3), the total power output may be written

$$\pi a^2 P_r = \frac{V_a^2}{\pi^2 \omega L} \tag{9.4}$$

The voltage swing V_a should be small enough to avoid the field in the intrinsic region being reduced to zero during the negative half-cycle. A safe value would be $V_a = V_d/2$, giving an efficiency of about 30%. In the negative half the field must not fall below about 5 kV/cm for the carrier velocity to remain independent of power, which requires that $V_d > 10^4 W$, with W in centimetres. To confine avalanching to the vicinity of the n^+-p junction requires that $V_d (=2V_a)$ shall be less than $0.4E_c W$, thus ensuring that the maximum field is always less than $0.6E_c$. Thus it is required that $10^4 W < V_d < 0.4EW$. For $W = 5 \times 10^{-4}$ cm the critical field would be about 400 kV and V_d would have to be less than 80 V. For $\omega L = 10\,\Omega$ the maximum power output would be about 16 W.

An upper limit on frequency is set by the fact that, for sufficiently narrow junctions, with breakdown voltages of 10 V or less, the current is generated by internal field emission (Zener effect) and not by multiplication. Although it was shown by Read that negative resistance operation on Zener current is possible, this mode of operation is much less effective.

Read's original paper discusses in detail the physics of multiplication, space charge and carrier flow, and derives approximate solutions governing

Avalanche Diode Oscillators

the behaviour of the diode, starting with Poisson's equation together with the equations of current continuity:

$$\frac{\partial p}{\partial t} = \frac{-1}{q}\frac{\partial I_p}{\partial x} + av(n + p)$$

$$\frac{\partial n}{\partial t} = \frac{1}{q}\frac{\partial I_n}{\partial x} + av(n + p) \qquad (9.5)$$

From the above equations it may be shown that, with small amplitudes of oscillation, for an applied periodic voltage with bias $E_0 = E_c + E_a \sin \omega t$,

$$\log_e \frac{I_0(t)}{I_0(0)} = \frac{2(m+1)}{\pi}\frac{\tau}{\tau_1}\frac{E_a}{E_c}\left(1 - \cos\frac{\pi t}{\tau}\right) \qquad (9.6)$$

In the above expression ω is written as the optimum frequency π/τ; τ_1 is the transit time across the multiplication region; and m is the exponent in the relationship $\alpha(E) \approx E^m$. The current/field relationships as functions of time are shown in Fig. 9.3.

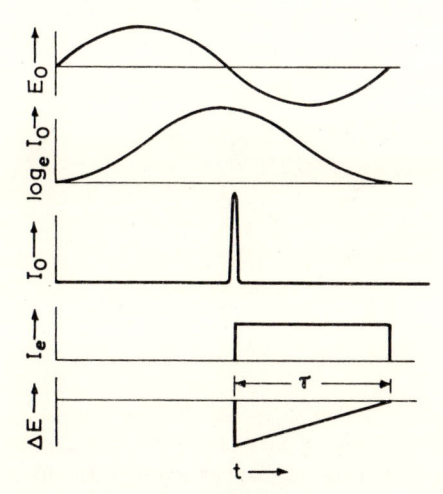

FIG 9.3 Current and field relationships for a sharp current pulse
(From Ref. 1)

It may be shown from eqn. (9.6) that for the small-signal case the Q-factor and admittances are given by

$$Q = \frac{\pi}{2}\left(1 - \frac{\pi^2 \tau_1}{2(m+1)I_d}\right)$$

Microwave Semiconductor Devices

and

$$Y_r = \frac{\pi Q}{1+Q^2}, \quad Y_i = \frac{\pi}{1+Q^{-2}}. \tag{9.7}$$

Gilden and Hines [11] have derived the following expression for the diode impedance under small-signal conditions:

$$Z = r_s + \frac{W^2}{v\epsilon A(1-\omega^2/\omega_a^2)} + \frac{1}{j\omega C(1-\omega_a^2/\omega^2)} \tag{9.8}$$

where r_s is the series resistance of the diode, C is the total depletion layer capacitance with the diode biased just below avalanche, and ω_a is the *avalanche frequency*, given by

$$\omega_a^2 = \frac{2\bar{\alpha}_a v I_0}{\epsilon}$$

$\bar{\alpha}_a$ being the derivative of the average ionization coefficient with respect to the electric field. The equivalent circuit is therefore as shown in Fig. 9.4.

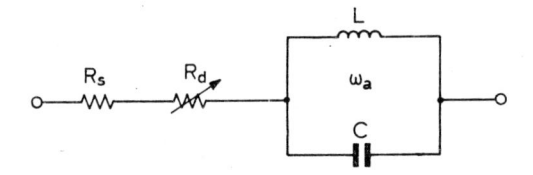

FIG 9.4 Avalanche diode equivalent circuit at small transit angle

$$C = \frac{\epsilon A}{I_d + I_a}$$

The efficiency is defined as the ratio of a.c. power P_r to the power P_d delivered from the d.c. source. When the source current is of the form of a sharp pulse the efficiency may be shown to be

$$\eta = \frac{-2}{\pi} \frac{V_a}{V_d} \cos \pi t_1 \tag{9.9}$$

where t_1 is the time at which the current pulse appears. For $V_a = V_d/2$ and $t_1 = 1$, then $\eta = 1/\pi$, i.e. the efficiency can exceed 30%.

9.3 p–n JUNCTION DIODE OSCILLATORS

The first avalanche diodes which were observed by Johnston *et al.* [2] to give microwave oscillations were of the form illustrated in Fig. 9.5. The voltage required for oscillation was found to correspond to an electric field of about 2 kV/cm in the 150 μm *n*-type region, a maximum output being reached at a certain voltage above threshold. With these diodes 80 mW of power was obtained at 12 GHz, representing an efficiency of

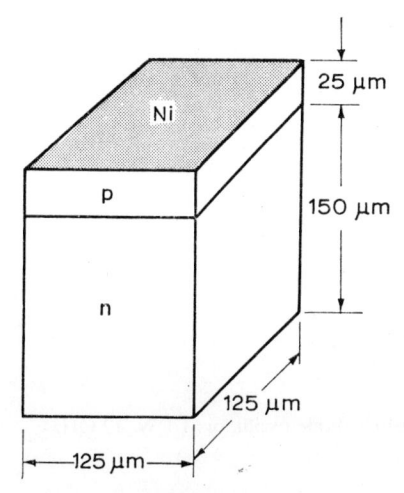

Fig 9.5 Early Bell Telephone Laboratories diode oscillator
(*From Ref.* 2)

0·5%. An approximately linear increase in output power with frequency was observed which was attributed to a redistribution of the alternating component of field between the space-charge depletion-layer capacitance and the a.c. impedance of the drift region.

It has been demonstrated by De Loach [15] that a simple *p–n* junction diode under reverse bias behaves as two Read diodes back to back. A disadvantage of this type of oscillator compared with the Read structure is the need to pass substantial currents (up to 2000 A/cm^2) in order to establish the required field in the undepleted region. This causes a loss of d.c. power and severely limits the efficiency. It also tends to make such oscillators more noisy. However, the need to operate well into the breakdown region means that dislocations in the semiconductor material are of small

Microwave Semiconductor Devices

importance and the manufacturing problems relatively simple. Since electron velocity is not entirely independent of field in the drift region, strong electronic tuning is possible. Parametric oscillations, especially lower sidebands, can be generated in varactor-like structures by the large r.f. fields of the fundamental [16].

The p–n junction oscillators have the highest power capability amongst all solid-state devices at the present time. For high values of mean power, careful attention must be paid to heat sinking and internal thermal resis-

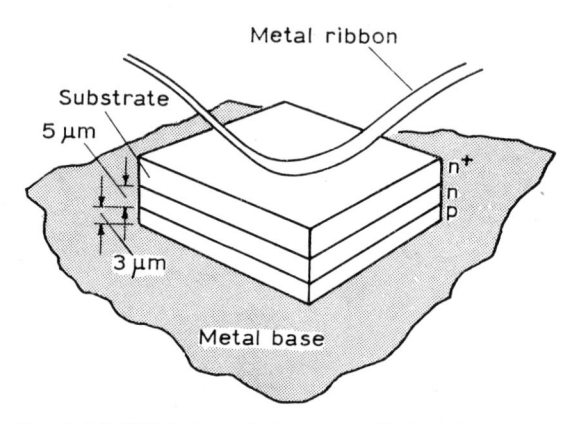

Fig 9.6 Bell Telephone Laboratories diode oscillator: 1·1 W, 12 GHz
(*From Ref.* 33)

tance. The first diode to give c.w. oscillations in excess of 1 W in the 3 cm band was developed by Misawa at the Bell Telephone Laboratories and is illustrated in Fig. 9.6. This diode is mounted "upside down" compared with conventional arrangements, so that the heat has only to pass through the p-layer (about 3 μm thick) thereby facilitating low thermal resistance, further enhanced by a tight control of the n-layer profile with minimal thickness of about 5 μm. Further refinements at the same laboratories, including a diamond heat sink, resulted in a diode giving 4·7 W at 13·3 GHz, previously indicated [8]. It is currently thought that 10 W is attainable with a copper substrate and 15 W with a diamond heat sink [17].

9.4 p–i–n DIODE OSCILLATORS

Microwave oscillations can also be obtained from a p–i–n structure, as first described by Misawa and Marinaccio [18]. Such diodes operate at extremely high current densities but can give the best efficiencies. They are generally easier to fabricate than devices having the Read structure but are more difficult than the simple *p–n* junction diodes. Likewise their power output and noise performance tend to be intermediate between those of the two other types.

The avalanche characteristics of *p–i–n* diodes have been studied by Egawa [19] in a different context, and Giblin [20] has published preliminary results of a theoretical investigation aimed at elucidating the operation of such devices as avalanche oscillators. It may be shown that at sufficiently high current densities the static reverse characteristics of

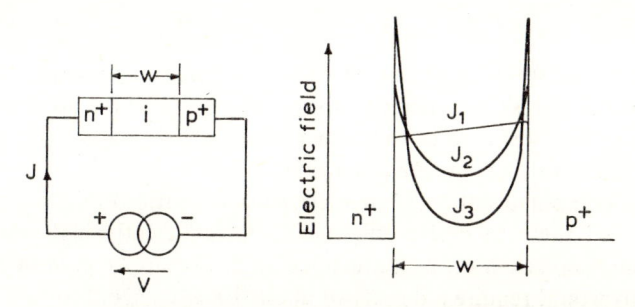

FIG 9.7 Electric field v. current density for *p–i–n* avalanche diode
(*From Ref. 20: reproduced by permission of the Services Electronics Research Laboratory (Ministry of Defence, Navy)*)

p–i–n diodes exhibit a current-controlled negative resistance, the same characteristics being preserved up to frequencies in the microwave range, i.e. until the transit time for carriers across the *i*-region becomes a significant fraction of the r.f. cycle. For this condition to apply at a frequency of 10 GHz the width would have to be about 1 μm.

The physical origin of the characteristic in terms of voltage and current density may be understood with reference to Fig. 9.7. If it is assumed that the holes and electrons have the same ionization coefficient and also move with the same saturation drift velocity v_s, Poisson's equation becomes

$$v_s \epsilon \frac{dE}{dx} = qnv_s + J_T - J_n \qquad (9.10)$$

where q is the modulus of electronic charge, n is the net impurity doping in the *i*-region, and J_p, J_n are the hole and electron current densities, whose sum is J, the total current density.

At breakdown, when the current density is still small compared with the fixed charge density, the field is represented by the curve labelled J_1 (the corresponding current density). Close to the n^+-i boundary it is approximately true that $J_n = J$, and similarly at the $i-p^+$ boundary that $J_p = J$. It follows that when $J \gg qnv_s$ the field near to each boundary must dip towards the centre, the dip increasing as J rises. Thus curves J_1, J_2, J_3 represent the field at increasing values of J. There is an increase in field at the boundaries which is explained as follows.

At breakdown ($J = J_1$), when each electron-hole pair generated produces on average one other pair,

$$\int \alpha \, dx = 1 \tag{9.11}$$

where α is the ionization coefficient. Since α is in increasing function of the field, the lower this dips the higher must the peaks become at the boundaries. There is also some decrease in the area under the field curves (i.e. total voltage) at sufficiently high current densities.

Giblin has computed static V/J characteristics for various thicknesses of diode and used these to derive efficiencies. At sufficiently high current densities values in excess of 50% are calculated. For an i-region of 2μm thickness this would require a density of about 10^7A/cm^2, but for a 50μm layer the corresponding figure would be about $2 \times 10^4 \text{A/cm}^2$.

Whether or not Giblin's analysis holds the key to an explanation of anomalous diodes seems rather uncertain. The maximum efficiency of 60% indicated in Table 9.1, for instance, is only consistent with his results if it is assumed that the effective current density was about 1000 times greater than that estimated from the total current and cross-sectional area. This is conceivable if the current does in fact concentrate into one or more filamentary paths. Alternatively, work at the Bell Telephone Laboratories [21] has suggested that the ultra-high frequencies of anomalous diodes are associated with multi-frequency large-amplitude terminal waveforms which modulate the avalanche multiplication and cause non-saturation effects of the carrier drift velocities to become important.

9.5 NOISE PERFORMANCE

A noise theory for avalanche diodes has been derived by Hines [22] with particular reference to amplifiers. Assuming that the noise performance can be described in terms of a noise current of mean-square effective value

Avalanche Diode Oscillators

I_n^2 operating into a oad resistance R_L, the noise figure is given by the general expression

$$F = 1 + \frac{I_n^2 R_L}{G k T_0 B} \qquad (9.12)$$

The theory depends on finding the various components of the noise current I_n, the first of which, i_{n0}, may be taken as a fictitious idealized zero alternating current which would flow in the avalanche if it were possible to maintain the electric field constant at exactly the critical value E_0 which is necessary to give a steady avalanche. Secondly, a component i_{ne} is considered to be induced in the avalanche zone and to correlate with i_{n0}. The third component is the displacement current in the avalanche zone, $j\omega \varepsilon A E_a$, so that the total current is given by

$$I_n = i_{n0} + i_{ne} + j\omega \bar{\varepsilon} A E_a \qquad (9.13)$$

Using the above basis, Hines derived an expression for the amplifier noise figure at high gain:

$$F = 1 + \frac{1}{4 m \omega^2 \tau_x^2} \frac{q V_a / kT}{1 - (\omega_a/\omega)^2} \qquad (9.14)$$

where V_a = Voltage drop across avalanche zone
τ_x = Statistical variation of collision interval
ω_a = Avalanche frequency
m = Exponent, assuming that the ionization coefficient varies as the mth power of the electric field.

The following are typical values for the various terms in eqn. (9.14):

$V_a = 6\text{V} \qquad \tau_x = 5 \times 10^{-13}\text{s}$
$m = 6 \qquad \omega_a/\omega = 0\cdot 5$

At a frequency of 10 GHz the above values give a predicted noise figure of some 40 dB, which could be even larger at high signal level, when the fluctuation in peak current between cycles may be larger. Thus the relatively high noise level observed with avalanche oscillators appears to be basic in origin, although many useful applications can nevertheless be found. In terms of practical results, the figure derived above is typical of values quoted for silicon oscillators but is an order of magnitude higher than has been obtained with germanium and gallium-arsenide devices [13].

9.6 COMPOSITE STRUCTURES

Bell Telephone Laboratories have reported on the possibilities for operating several avalanche diodes in parallel, when they are spaced sufficiently closely to operate as a single oscillator but with essentially individual heat sinking [23]. It was found that the efficiency improved approximately as the diode diameter and the power capability increased, and not unexpectedly, as the number of wafers.

9.7 STATE OF THE ART

Table 9.1 (page 186) on the state of the art with avalanche transit-time diode oscillators was prepared in February, 1970.

9.8 CONSTRUCTION TECHNIQUES

9.8.1 Read Diodes

One possible way of providing the Read-type structure is to employ high-resistivity bulk material and apply a triple-diffusion process. With silicon,

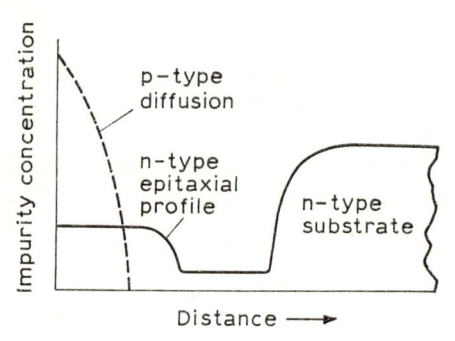

FIG 9.8 Typical impurity profile of epitaxial Read diode

for instance, bulk n-type material may be used having a resistivity of several thousand ohm-centimetres. A phosphorus diffusion may then give an n^+-type rear contact, and a low surface-concentration phosphorus diffusion, followed by a "drive-in" heat cycle, may be used to give the n-type step. Finally, the p^+-profile may be provided through a boron diffusion.

It is also possible to make such structures using epitaxial material in which the rear contact is provided at the substrate, but with the double front-end diffusions exactly the same. There will now be a rear concentration profile arising from out-diffusion at the substrate and redeposited impurities, as discussed in Chapter 3. A typical composite profile obtained in the way is illustrated in Fig. 9.8.

9.8.2 p–n Diodes

p–n avalanche diodes may also be prepared in the same general manner as the Read diodes described above, but with a uniform impurity concentration in the epitaxial layer, which implies that the diffusion depth is not so

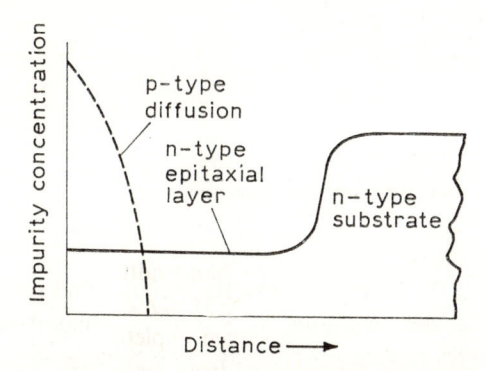

Fig 9.9 Pattern of impurity profile for *p–n* diode oscillator

critical; but it is desirable that voltage breakdown should occur before the edge of the depletion layer reaches the out-diffusion profile, so as to avoid the higher dislocation density which usually occurs there and which may precipitate burnout. The pattern of impurity profile would be as illustrated in Fig. 9.9, the associated field being of triangular form.

The *p–n* junction diodes from which microwave oscillations were first reported by Johnston *et al.* were of *n*-type bulk material into which boron was diffused on one face, the thickness being controlled by lapping. The subsequent processing was to apply electroless nickel to both faces and to sinter at 800°C for 5 min in a nitrogen atmosphere, followed by replating with nickel and then with gold. Sectioning into wafers was then carried out ultrasonically, followed by a 10 s etch in CP-8 to remove cutting damage (3 parts HNO_3, 1 part HF). In subsequent work on germanium avalanche devices at the Bell Telephone Laboratories, pellet separation has been achieved by etching, using gold contacts as the masking material.

Microwave Semiconductor Devices

9.8.3 Other Structures

A particular planar structure which is reported to have a noise performance much superior to conventional planar device is illustrated in Fig. 9.10 [10]. This is achieved by diffusing boron through an oxide mask into an n-type epitaxial layer of some 10 Ω-cm resistivity on an n^+-substrate. Diffusion of the boron is made sufficiently deep for the breakdown voltage not to be influenced by the curvature of the junction. Breakdown then

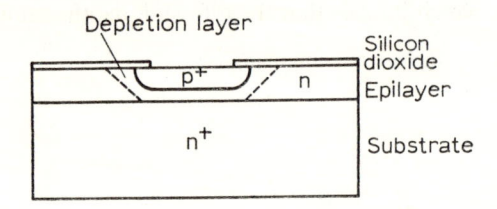

FIG 9.10 Planar avalanche diode with restricted depletion layer
(*From Ref. 10*)

occurs only in the flat part of the junction, and a bandwidth of 1 kHz has been reported for such a device operating at 5·3 GHz compared with 50–100 GHz for planar diodes having no restricted depletion layer. The figure of 1 kHz is comparable with that obtained from mesa structures.

The anomalous diodes first reported by RCA Laboratories employed a somewhat similar p^+–n–n^+ structure with an n-region having a resistivity of about 5Ω-cm and 8–10 μm in width.

REFERENCES

1 READ, W. T., "A proposed high-frequency negative-resistance diode", *Bell System Tech. J.*, **37**, p. 401 (1958).

2 JOHNSTON, R. L., DE LOACH, B. C., and COHEN, B. G., "A silicon diode microwave oscillator", *ibid.*, **44**, p. 369 (1965).

3 LEE, C. A., BATDORF, R. L., WIEGMANN, W., and KAMINSKY, G., "The Read diode—an avalancing transit-time, negative-resistance oscillator", *Appl. Phys. Letters*, **6**, p. 89 (1965).

4 BURRUS, C. A., "Millimeter-wave oscillations from avalanching p–n junctions in silicon", *Proc. IEEE*, **53**, p. 1257 (1965).

5 NAPOLI, L. S., and IKOLA, R. J., "An avalanching silicon diode microwave amplifier", *ibid.*, **53**, p. 1231 (1965).

Avalanche Diode Oscillators

6 PRAGER, H. J., CHANG, K. K. N., and WEISBROD, S., "High-power high-efficiency silicon avalanche diodes at ultra-high frequencies", *Proc. IEEE*, **55**, p. 586 (1967).

7 CHANG, K. K. N., "Avalanche diodes as UHF and L-band sources", *RCA Rev.*, **30**, No. 1, p. 3 (1969).

8 SWAN, C. B., "Improved performance of silicon avalanche oscillators mounted on diamond heat sinks", *ibid.*, **55**, p. 1617 (1967).

9 GILDEN, M., and MORONEY, W., "High power pulsed avalanche diode oscillators for microwave frequencies", *ibid.*, **55**, p. 1227 (1967).

10 KOCK, H. G., DE NOBEL, D., VLAARDINGERBROEK, M. T., and DE WAARD, P. J., "Continuous-wave planar avalanche diode with restricted depletion layer", *ibid.*, **56**, p. 105 (1968).

11 GILDEN, M., and HINES, M. E., "Electronic tuning effects in the Read microwave avalanche diode", *Trans. IEEE*, **ED-13** p. 169 (1966).

12 MIDFORD, T. A., "Effects of uniaxial stress on avalanche transit time diode oscillators", I.E.E.E. International Electron Devices Meeting, Washington, D.C. (1966).

13 KILPATRICK, T. H., "Avalanche diodes top 400-W pulse at 1 GHz", *Microwaves*, **6**, No. 6, p. 10 (1967).

14 MCKAY, K. G., "Avalanche breakdown in silicon", *Phys. Rev.*, **94**, p. 877 (1954).

15 DE LOACH, B. C., "Avalanche transit-time microwave oscillators", *Trans. IEEE*, **ED-13**, p. 181 (1966).

16 IRVIN, J. C., "Gallium-arsenide microwave oscillators", *Trans. IEEE*, **ED-13**, p. 208 (1966).

17 FITZSIMMONS, G. W., "Microwave generation with avalanche diodes and two-valley Gunn and LSA mode devices", *Microwave J.*, **11**, No. 2, p. 45 (1968).

18 MISAWA, T., and MARINACCIO, L. P., "A $\frac{1}{4}$ W silicon p–i–n X-band IMPATT (Impact Avalanche Transit Time) diode", *Bell System Tech. J.*, **45**, p. 989 (1966).

19 EGAWA, H., "Avalanche characteristics and failure mechanism of high-voltage diodes", *Trans. IEEE*, **ED-13**, p. 754 (1966).

20 GIBLIN, R. A., "High-efficiency operation of avalanche-diode oscillators", *Electronics Letters*, No. 3, **4**, p. 52 (1968).

21 JOHNSTON, R. L., SCHARFETTER, D. L., and BARTELINK, D. J., "High-efficiency oscillations in germanium avalanche diodes below transit-time frequency", *Proc. IEEE*, **56**, p. 1611 (1968).

22 HINES, M. E., "Noise theory for the Read-type avalanche diode", *Trans. IEEE*, **ED-13**, p. 158 (1966).

23 SWAN, C. B., MISAWA, T., and MARINACCIO, L., "Composite avalanche diode structures for increased power capacity", *ibid.*, **ED-14**, p. 584 (1967).

24 IGLESIAS, D. E., and EVANS, W. J., "High-efficiency c.w. IMPATT operation", *Proc. IEEE*, **56**, p. 1610 (1968).

25 IGLESIAS, D. E., "Circuit for testing high-efficiency IMPATT diodes", *Proc. IEEE*, **55**, p. 2065 (1967).

26 MELICK, D. R., "High-frequency pulsed GaAs avalanche diodes", *ibid.*, p. 435.
27 RULISON, R. L., GIBBONS, G., and JOSENHANS, J. G., "Improved performance of IMPATT diodes fabricated from Ge", *ibid.*, p. 223.
28 KIM, C. K., ARMSTRONG, L. D., and MATTHEI, W. G., "High-power Schottky-barrier avalanche diodes". Paper presented at IEEE Devices Research Conference, Rochester University, N.Y., June, 1969.
29 SWAN, C. B., MISAWA, T., and MARINACCIO, L., "Composite avalanche diode structure for increased power capacity", *Trans. IEEE*, **ED-14**, p. 584 (1967).
30 LIU, S. G., "GaAs avalanche microwave oscillator with 1-W power output", *Proc. IEEE*, **55**, p. 689 (1967).
31 IRWIN, J. C., "GaAs avalanche microwave oscillators", *Trans. IEEE*, **ED-13**, p. 208 (1966).
32 EDWARDS, R., *et al.*, "Millimeter-wave silicon IMPATT diodes". Paper presented at IEEE Conference on Electron Devices, Washington, D.C., 1969 (Proceedings to be published.)
33 *Microwaves*, **5**, No. 12, p. 16 (1966).

10

Integrated Circuits

10.1 INTRODUCTION

Microwave integrated circuits are still in their infancy. Their real attraction commercially depends on significant consumer applications arising which are based on the low-cost high-volume manufacturing techniques developed for lower frequencies and also on the conversion to solid-state phased-array radars from conventional tube-dish techniques. There are also technical improvements which integration of solid-state devices and strip-line

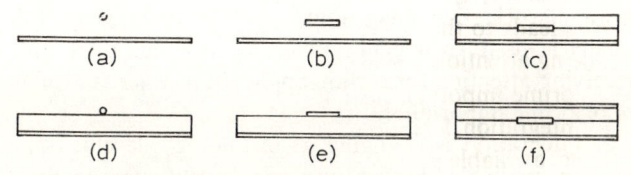

FIG 10.1 Various forms of strip transmission line
(a) Cylindrical conductor, air-spacing
(b) Strip conductor, air-spacing
(c) Strip conductor, bedded in dielectric, single ground plane
(d) Cylindrical conductor, dielectric spacing
(e) Strip conductor, dielectric spacing (microstrip)
(f) Strip conductor, dielectric spacing between two ground planes (triplate)
(*From Ref.* 26)

circuitry can offer apart from the above considerations and which justify work which is not aimed at quantity production. Airborne applications invariably emphasize the small size/weight aspect, but this may be accomplished by miniaturization and does not necessarily involve integration. One important benefit which does stem from integration, however, is increased reliability, a second desirable feature being the elimination of connectors between many components, of particular importance in applications involving broadband filters.

Integration implies the use of strip transmission lines. Several alternative forms are illustrated in Fig. 10.1, of which principally two have found

203

practical application, i.e. triplate and microstrip ((f) and (e)). The latter form, because of its inherent simplicity, is receiving by far the widest attention and we shall imply the use of microstrip throughout this chapter.

There are two basic constructional approaches—monolithic and hybrid. In the monolithic approach, high-resistivity semiconductor material is used as the substrate, into which junction devices are formed by planar techniques, subsequent interconnection being effected with deposited metallic strip conductors, which are also employed as transmission-line circuit-elements. In hybrid systems an insulator is used as the substrate, e.g. high-alumina ceramic or sapphire. Strip conductors are then deposited to form the appropriate circuits, leaving series gaps or holes to the substrate, for shunt connection, into which separately mounted semiconductor devices or passive elements may be inserted.

Microelectronics generally is concerned with thick-film and thin-film deposition techniques, the former implying deposition of conductors, resistors and capacitors by a printing process, and the latter implying some form of vacuum deposition, to be followed by plating, etching, oxidation, etc.

For microwave work, the dimensional control obtained by thin-film processes usually leads to their being preferred, although thick-film techniques are receiving attention for certain applications where flexibility and economy are of prime importance.

The level of integration is changing as technology develops and new materials become available. Device and component manufacturers are being forced into much closer liaison with each other and devices are required in new forms. Integrated circuits give scope for absorbing the stray reactances associated with devices into the circuit-elements, thereby achieving better overall performance. New measurement problems arise because it is often not appropriate to measure as discrete units devices designed specifically for use in integrated circuits. The best way of evaluating such devices is often by examination of the completed circuit or subassembly. Circuits which have so far been successfully integrated into subsystems include mixers, switches, sources, circulators, amplifiers and limiters.

10.2 PROPERTIES OF MICROSTRIP TRANSMISSION LINE

It is important that in a chapter dealing with microwave integrated circuits we should consider briefly the properties of microstrip. Assadourian and

Integrated Circuits

Rimai [1] are credited with the first theoretical investigation, followed by Dukes [2], who used an electrolytic tank in his studies. More recently, Wheeler [3] has derived key equations for parallel-strip transmission line

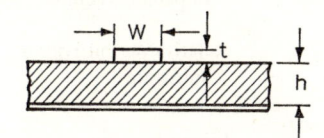

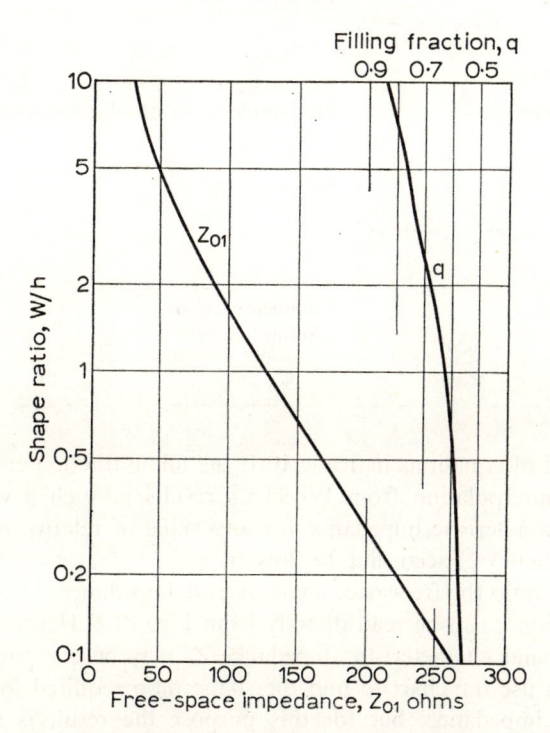

FIG 10.2 Free-space impedance and filling fraction for microstrip
(From Ref. 5)

using conformal mapping techniques, these equations having been subsequently verified experimentally [4]. Taking Wheeler's results, Presser [5] has given a general-purpose design chart which is useful over a wide range of shape ratios (i.e. strip width to dielectric thickness) for any substrate of known permittivity. This plot is reproduced in Fig. 10.2, together with

205

TABLE 10.1

Characteristic impedance (ohms) $Z_0 = \dfrac{Z_{01}}{\sqrt{\epsilon_r'}}$	Free-space characteristic impedance (ohms) Z_{01} Relative permittivity of substrate ϵ_r
Effective relative permittivity $\epsilon_r' = 1 + q(\epsilon_r - 1)$	Filling fraction q
Wavelength in transmission line $= \dfrac{\lambda_0}{\sqrt{\epsilon_r'}}$	Free-space wavelength λ_0 Dissipation factor of substrate $\tan \delta$
Dielectric losses (nepers/unit length) $\alpha_d = \dfrac{q\pi}{\lambda_0} \dfrac{\epsilon_r}{\sqrt{\epsilon_r'}} \tan \delta$	Frequency (GHz) f
Conductor losses (nepers/unit length) $\alpha_c = \dfrac{62 \cdot 8}{Z_0 W} (f\rho)^{1/2}$	Conductor resistivity (ohm-metres) ρ

an accompanying set of equations in Table 10.1, the filling fraction curve being obtained by interpolation from Wheeler's results in such a way that the error in characteristic impedance for any value of relative permittivity, ϵ_r, is less than 3%, according to Presser.

For a given shape ratio the free-space characteristic impedance Z_{01} and effective filling fraction q can be read directly from Fig. 10.2. Hence the relative permittivity and characteristic impedance Z_0 may be computed. It is also possible to use the chart to find the shape ratio required for a given characteristic impedance, but for this purpose the result is not obtained as easily and an iterative procedure is suggested.

If the strip thickness t is appreciable, a width correction factor must be added to the actual width, given by

$$\Delta W = \frac{t}{\pi}\left(1 + \frac{2h}{t}\right) \qquad (10.1)$$

where h is the substrate thickness.

Integrated Circuits

The above formula applies for $\Delta W/t > 1\cdot 33$ for moderate values of permittivity.

A further important property associated with transmission-line resonators, particularly useful in estimating attenuation coefficients, is the unloaded Q-factor, given in terms of the dielectric and conductor Q-factors, Q_d and Q_c, by

$$Q_0 = \frac{Q_d Q_c}{Q_d + Q_c} \qquad (10.2)$$

where, for microstrip,

$$Q_d = \frac{\epsilon_r'}{q\epsilon_r} \frac{1}{\tan \delta} \qquad (10.3)$$

and

$$Q_e = \frac{Z_{01} W}{6} \sqrt{\frac{f}{p}} \qquad (10.4)$$

It has been found that the filling fraction q is not strictly independent of frequency so that Fig. 10.2 is valid only at the lower microwave frequencies. Some early results on this frequency dependence were presented by Hartwig *et al.* [6] in 1968. Results showing close agreement between a theoretical treatment and experimental values up to 10 GHz were subsequently given by Zysman and Varon [7]. Recent work on ring resonators by Troughton [8] has shown the effective relative permittivity to rise steadily with frequency up to at least 18 GHz, at which point values were found to be some 5–25% greater than at low frequencies. For a given low-frequency value of Z_0 the increase was shown to be a steadily rising function of dielectric thickness. This work involved alumina microstrip with Z_0 in the range 25–75 Ω and thickness 0·25–1·25 mm.

10.3 HYBRID CIRCUITS

Where extremely low loss, good dimensional stability or high temperature capability is required, high-purity alumina-ceramic substrates have no equal in hybrid circuit work, with the possible exception of sapphire, which is usually excluded on account of either cost or anisotropy. For more general use, however, a range of strip-line laminates has become available with microwave relative permittivities ranging from 2·1 to 2·7 without

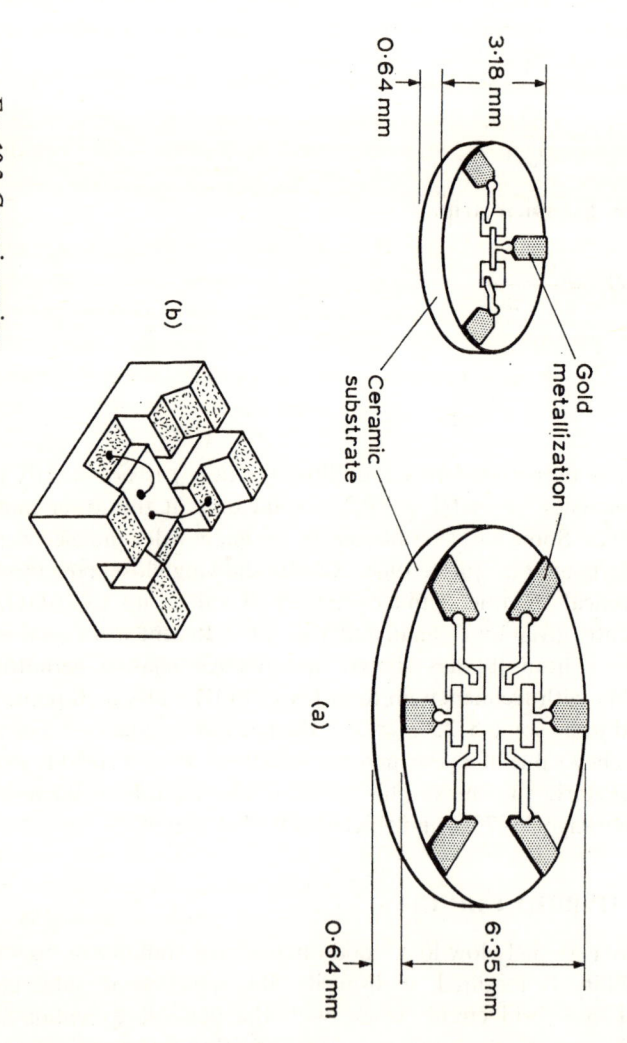

FIG 10.3 Ceramic carriers
(a) Beam-lead devices *(From Ref. 27)*
(b) Leadless inverted device *(Frenchtown/CFI)*

Integrated Circuits

ceramic loading, and from about 3 to 25 for ceramic-filled plastic laminates, compared with about 6·5 for alumina ceramics and 3·5 for sapphire [9]. The choice of material depends on the particular application, bearing in mind, besides permittivity, such features as dissipation factor, surface finish, homogeniety, isotropy, dimensional stability, temperature range, machinability and other physical features.

Glazed ceramic of 96–97% alumina content is a material particularly suitable for low-loss microwave circuits, provided the surface roughness does not exceed 0·25 μm, which may be facilitated by polishing. Gold conductors, deposited by thin-film techniques, offer the best long-term stability and resistance to chemical attack. A vacuum deposition method which has been found very satisfactory is first to clean the surfaces by ionic bombardment in a glow discharge, followed by "flash" of nichrome (say 1·5 × 10^{-3} μm in thickness) and then a few hundredths of a micron of gold, subsequently sintered to improve adhesion. The final thickness of gold (up to a few microns) is attained by electroplating, the thickness being made two or three times greater than the skin depth at the appropriate microwave frequency.

The conductor pattern is obtained by photo-etching. To this end a mask may be prepared by drawing the circuit to a 10:1 scale on a Rubylith film and generating the photomask by a 10:1 photographic reduction, larger ratios sometimes being necessary. The gold layer on the side upon which the conductor pattern is to be formed is coated with photoresist and exposed through the prepared mask, the exposed metal part being then removed by etching, the substrate also being protected during this operation. Active and passive devices are then bonded in chip form to the substrate.

Examples of chip arrangements in current use are shown in Fig. 10.3. One example is a *beam-lead* structure built on a Schottky-barrier diode chip. (Beam-lead devices have the surrounding semiconductor material etched away after the conductors have been deposited, and such techniques have also been applied to resistors and capacitors.) The other example is that of a "leadless inverted device" (LID), which is essentially a form of ceramic housing and may contain any one of a number of components. Techniques of bonding include thermal compression, ultrasonic and thermal pulse bonding, of which the latter is particularly suitable for beam-lead devices with almost any substrate material of interest, being additionally relatively insensitive to contamination and having wide bonding margins [10].

Microwave Semiconductor Devices

With thick-film hybrid circuits, after completion of the artwork, the pattern is transferred to a screen using techniques such as transfer emulsion, direct emulsion, metal foil or etched screen [11]. The pattern is then printed from the screen on to the substrate, followed by drying and firing schedules. As well as conductors, resistors (10 Ω to 10 MΩ) can be prepared by this technique, and also capacitors, with single-layered units having capacitance values up $0.03\,\mu\text{F/cm}^2$.

10.4 MONOLITHIC CIRCUITS

The possibility of using semiconductor material as the dielectric for monolithic microwave integrated circuits has received considerable attention. The forming of devices integrally with the substrate is attractive from the point of view of simplicity, cheapness and improved performance. Silicon and gallium arsenide are two materials which have been employed to date, but most of the work has been with silicon, which can be obtained as the purest and most uniform dielectric material currently available and whose surface may be polished to any desired degree.

The resistivity and permittivity of silicon have been shown to be the same at microwave frequencies as at low frequencies [12]. It has been reported by Hyltin [13] that the line losses with silicon are predominantly ohmic, which implies constant loss per unit length and a loss per wavelength which decreases with increasing frequency. The dissipation factor is in fact given by

$$D = \frac{1}{\omega \rho E} \tag{10.5}$$

The curve of line loss per centimetre as a function of silicon resistivity is reproduced from Hyltin's paper in Fig. 10.4. In practice, resistivities of over 3000 Ω-cm can be achieved fairly readily, but care must be taken in subsequent processing if such values are to be maintained in the final circuit. Results from the same paper for measurements of the characteristic impedance of microstrip line on silicon dielectric are reproduced in Fig. 10.5. It was further shown by plotting line loss as a function of conductor thickness, using aluminium strip, that there is little advantage in building up beyond $3.5\,\mu$m. Also, the line loss does not increase significantly with temperatures up to about 110°C for 1500 Ω-cm silicon.

The problems encountered in preparing monolithic microwave circuits are, however, severe. For example, consider the requirements for a monolithic *p–i–n* modulator, with the modulating diode formed across the centre

Integrated Circuits

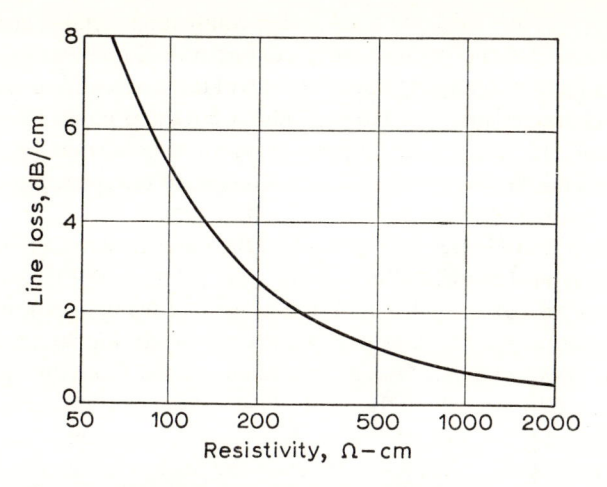

Fig 10.4 Microstrip loss *v.* dielectric resistivity
(*From Ref.* 13)

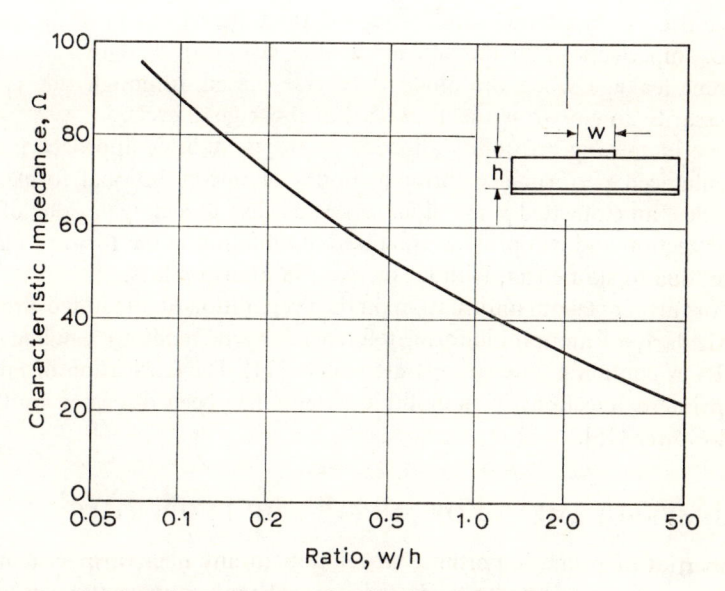

Fig 10.5 Impedance of microstrip with silicon dielectric
(*From Ref.* 13)

of a silicon slice, 0·25 mm thick, covered with a conducting metallic ground plane on the one side and carrying a strip conductor 0·25 mm wide on the other, these conductors contacting opposite faces of the diode. Such a system would have a characteristic impedance of about 50 Ω and could form an extremely broadband modulator. To prevent injection of carriers into the substrate at high microwave voltages it is customary to interpose layers of silicon dioxide between the silicon and the conductors.

It is all too easy to find after processing that the minority carrier lifetime of the bulk silicon has become unduly short ($\ll 1\,\mu s$). To prevent this, some form of gettering process is used, a common procedure being to expose the slice to an atmosphere containing phosphorus pentoxide for 20 min at a temperature of about 1050°C. Similar treatment, or the same operation, may also be used to provide diffused n^+-layers, boron diffusion being employed principally for the p^+-layers. It is found, however, that the phosphorus process can lead to adhesion problems with conductors which are subsequently deposited and also any silicon dioxide present is liable to be attacked. The degree of attack which can be tolerated is much less than with the preparation of a number of devices on a single slice.

A further problem commonly encountered is that of high contact resistance between deposited conductors and the diffused region of the diode. Also, imperfections in the silicon dioxide layer can lead to substantial current leakage when the diode is reverse biased, although this is not necessarily an embarrassment, at least at low signal levels.

One interesting possibility afforded by the monolithic approach is that of enhanced performance through diodes of unconventional form. For instance, an elongated p–i–n diode is just as easy to arrange as one of circular section and can provide enhanced attenuation in the forward biased state, due to skin effect, without increase in insertion loss.

For further general information on the preparation of integrated circuits, particularly of monolithic form, reference may be made to a publication by RCA containing fifteen separate papers [14]. The fabrication of microstrip interconnections in monolithic circuits has been discussed by Cash and Gower [15].

10.5 MEASUREMENTS OF MICROSTRIP PROPERTIES

Properties of prime importance in relation to any microstrip system are the characteristic impedance Z_0, guide wavelength λ_g and attenuation coefficient α. It is desirable to establish these properties experimentally on any new transmission system where high accuracy is required, since it is difficult to predict the effects of fringe fields, finite conductor size, radiation losses and the dielectric discontinuity at the insulator boundary.

10.5.1 Determination of Z_0

The simplified method for determining Z_0 is by means of a slotted line (or network analyser) in conjunction with a straight length of the microstrip terminated in a matched load. Coaxial/microstrip transitions can have v.s.w.r.s not exceeding 1·05 over an octave bandwidth (at least near the 10 cm band), implying that values of characteristic impedance of around 50 Ω can be measured to an accuracy of the order of ±2·5 Ω.

10.5.2 Determination of λ_g

The most direct method of measuring λ_g is by means of a travelling probe carriage, which may conveniently be constructed from a conventional waveguide standing-wave carriage modified to carry the dielectric probe at a small distance above the surface of the strip conductor and thereby sampling the fringe field. Alternatively λ_g may be determined by measuring the change in position of the minimum of the standing wave pattern as the distance to a short-circuit plane terminating the line is varied.

10.5.3 Measurement of α

The most satisfactory method of measuring the attenuation coefficient for low-loss substrates is from the Q-factor of a resonant section of the line, although, since resonant microstrip circuits lead to enhanced radiation, care must be taken with screening. Coupling is made to a section of line, which may be λ_g or perhaps $1·5\lambda_g$ in length, via the fringe capacitance of gaps at each end of the central section, with transitions from the outer sections to coaxial line and thence to rectangular waveguide. Such a system behaves as a double-output coupled cavity, the attenuation coefficient being given in terms of the internal Q-factor, Q_0, by

$$\alpha = \frac{\pi}{Q_0 \lambda_g} \qquad (10.6)$$

It is difficult to measure Q_0 directly from, say, a simple swept frequency display, since, if the gaps are made small enough to give adequate coupling for this purpose, it is found that the wanted response becomes swamped by unwanted cross-coupling. One effective way of obtaining Q_0 is by determining the minimum insertion loss, L_0, as the frequency is varied and noting the loaded Q-factor, Q_L, indicated from the frequency change associated with a 3 dB increase in insertion loss over the value L_0. Q_0 is then given by

$$L_0 = 20 \log_{10} \frac{Q_0}{Q_0 - Q_L} \tag{10.7}$$

If S_1 and S_2 are the v.s.w.r.s at resonance, looking in each of the two directions, it may be shown that

$$L_0 = 20 \log_{10} \frac{2(1 + S_1)(1 + S_2)}{S_2(1 + S_1) + S_1(1 + S_2)} \tag{10.8}$$

For silicon substrates, where the attenuation is of the order of 0·5 dB/cm, the attenuation coefficient can be found from the envelope of standing wave minima, using a travelling probe, but this method is not sensitive enough for substrates having lower loss.

10.5.4 Transitions to Microstrip

Simple transitions may be made to microstrip by such techniques as extending the inner conductor of a miniature coaxial connector to make connection via a tab [4]. The properties of transitions may be explored using techniques for lossless [16] or lossy [17] situations, as appropriate.

For a lossless transition to microstrip a useful method suggested by J. R. G. Twisleton, for obtaining the transition mismatch S_T and attenuation coefficient α, is to measure the variation of v.s.w.r. with frequency when the line is either open-circuited or short-circuited at its far end, when it may be shown that

$$S_T{}^2 = \frac{(S_1 S_0 - 1)(S_0 - S_2)}{(S_2 S_0 - 1)(S_0 - S_1)}$$

and

$$\tan 2\alpha L = \frac{(S_0 - S_1)(S_0 - S_2)}{(S_1 S_0 + 1)(S_2 S_0 + 1)} \tag{10.9}$$

where L is the line length, S_1 and S_2 are the maximum and minimum observed v.s.w.r.s, respectively, and S_0 is the measured v.s.w.r. with the tab short-circuited to the ground plane.

By using a computer it is possible to measure directly the properties of strip-line circuit components through transitions from waveguide or

Integrated Circuits

coaxial measuring systems. This can be done with on-line correction giving, for instance, a swept-frequency display of reflection coefficient for a microstrip load with transition errors removed [18,19].

10.6 APPLICATIONS

Probably the most obvious need for microwave integrated circuits is in phased-array radars, where large numbers of identical circuits are required.

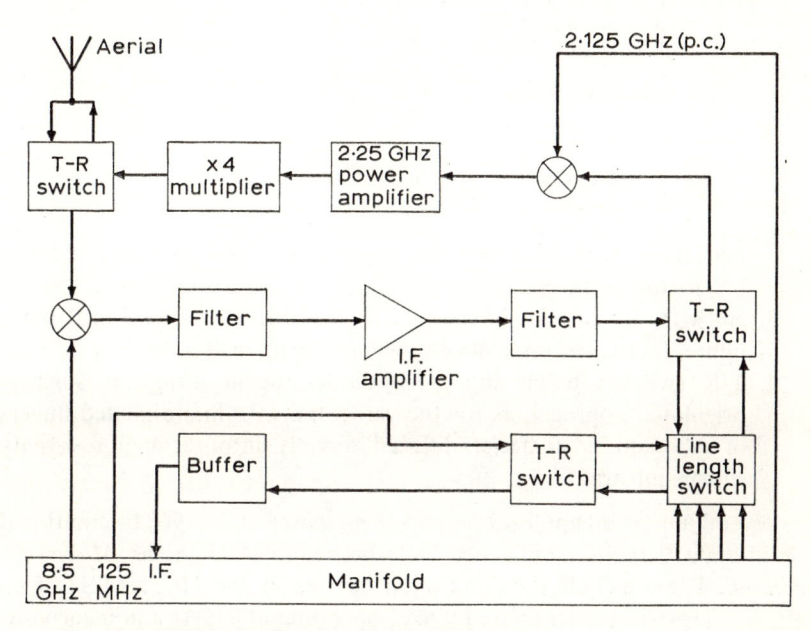

Fig 10.6 MERA circuit: microwave building unit
(Texas Instruments)

The first major system of this kind employing solid-state devices was the programme undertaken by Texas Instruments at Dallas, under T. M. Hyltin, for the project MERA (Molecular Electronics for Radar Applications) [20]. This 3 cm airborne radar system employed 600 1 W elements with a 100:1 pulse compression ratio, giving the equivalent of 600 kW peak power as used by a typical radar performing the functions of terrain following/terrain avoidance/ground mapping. A detailed diagram of the microwave building block is shown in Fig. 10.6.

Microwave Semiconductor Devices

The microwave semiconductor devices developed for integrating into this system include:

1. Surface-oriented Schottky-barrier mixers, formed in epitaxial pockets in high-resistivity silica substrates.
2. Frequency multipliers employing surface-oriented varactors, as illustrated in Fig. 10.7. These are formed in low-resistivity epitaxial

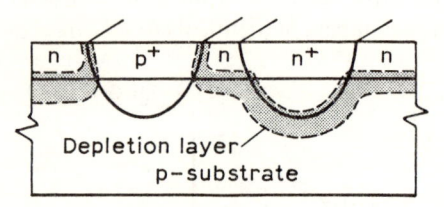

FIG 10.7 Surface oriented varactor
(Texas Instruments)

pockets in high-resistivity silicon, both n^+- and p^+-contacts being on the top surface.

3. 500 MHz i.f. pre-amplifiers with up to 20 dB gain and 3·6 dB noise figure, fabricated on a silicon bar measuring 0·30 in by 0·17 in.
4. T-R switches employing p–i–n diodes, again using the surface-orientated approach as for the varactors, with interdigitated fingers of p^+- and n^+-material diffused directly into the high-resistivity silicon substrate.

Several important applications papers presented at the 1968 International Solid-State Circuits Conference have been published in the *Microwave Journal*. These include the description by Napoli and Hughes [21] of an RCA microstrip tunnel-diode receiver operating at 9 GHz and incorporating a 4-port circulator fabricated on a ferrimagnetic substrate, the associated transmitter being an amplitude-modulated avalanche diode. Arnold et al. [22] describe a simple Doppler radar using a Gunn oscillator as the power source, the principle of the circuit being illustrated in Fig. 10.8. Kramer et al. [23] describe an integrated 3 cm c.w. radar front-end using an avalanche oscillator of 100 mW output, all except the source being constructed in ceramic microstrip. Botka et al. [24] describe front-end receiver developments also including a pulsed radar application.

Many established types of system are beginning to feel the impact of integration. A useful review of this activity in the United Kingdom has been prepared by Cornbleet [25], who discusses the development by various

Integrated Circuits

manufacturers of active devices which are compatible with hybrid/thin-film techniques employing chip or leadless inverted devices on ceramic microstrip. Mention is made of other types of diode developed for integrated circuits other than those already discussed, in particular planar backward-diode detectors by A.S.M. and $p-i-n$ modulators by A.E.I.

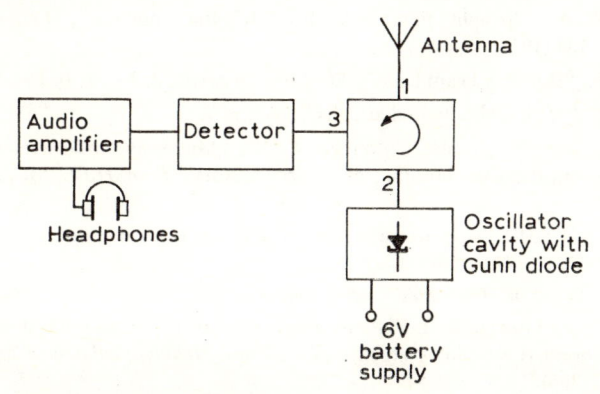

FIG 10.8 Gunn oscillator radar

We have indicated that numerous exploratory investigations are being pursued, but for the full potential growth of microwave integrated circuits to be realized, as well as tube-dish techniques giving way to solid-state phased-array designs, commercial and consumer applications must be developed, which seems inevitable as soon as technologies and costs have reached certain levels.

REFERENCES

1 ASSADOURIAN, F., and RIMAI, E., "Simplified theory of microstrip transmission systems", *Proc. Inst. Radio Engrs*, **40**, p. 1651 (1952).
2 DUKES, J. M. C., "An investigation into some fundamental properties of strip transmission lines with the aid of an electrolytic tank", *Proc. Instn Elect. Engrs*, **103B**, p. 319 (1956).
3 WHEELER, H. A., "Transmission line properties of parallel strips separated by a dielectric sheet", *Trans. IEEE*, **MTT-13**, p. 173 (1965).
4 CAULTON, M., HUGHES, J. J., and SOBEL, H., "Measurements on the properties of microstrip transmission lines for microwave integrated circuits", *RCA Rev.*, **27**, p. 377 (1966).
5 PRESSER, A., "R.F. properties of microstrip line", *Microwaves*, **7**, No. 3, p. 53 (1968)
6 HARTWIG, C. P., *et al.*, "Frequency-dependent behaviour of microstrip". Paper presented at G-MTT International Microwave Symposium, Detroit, Mich. (1968).

7 Zysman, G. I., and Varon, D., "Wave propagation in microstrip transmission lines". Paper presented at G-MTT International Microwave Symposium, Dallas, Texas (1969).

8 Troughton, P., "The evaluation of alumina substrates for use in microwave integrated circuits". Paper presented at 1969 European Microwave Conference, London (Proceedings to be pusblished).

9 Vossberg, W. A., "Stripping the mystery from strip-line laminates", *Microwaves* **7**, No. 1, p. 104 (1968).

10 Mallery, P., "Bonding beam leads", *Electron. Commun.*, **2**, No. 7, p. 15 (1968).

11 Early, R. C., "Thick-film hydoid circuits", *ibid.*, p. 7.

12 Nag, B. R., Roy, S. K., and Chatterji, C. K., "Microwave measurement of conductivity and dielectric constant of semiconductors", *Proc. IEEE*, **51**, p. 962 (1963).

13 Hyltin, T. M., "Microstrip transmission on semiconductor dielectrics", *Trans. IEEE*, **MTT-13**, p. 777 (1965).

14 Lohman, R. D., et al., *Integrated Circuits* (RCA, 1964).

15 Cash, J. H., and Gower, R. L., "Fabrication of microstrip interconnections for semiconductor microwave integrated circuits", *Trans. Metallurgical Soc. of AIME*, **236**, p. 388 (1966).

16 Shurmer, H. V., "Transformation of the Smith chart through lossless junctions", *Proc. Instn Elect. Engrs*, **105C**, p. 177 (1958).

17 Deschamps, G. A., "Determination of reflection coefficients and insertion loss of a waveguide junction", *J. Appl. Phys.*, **24**, p. 1046 (1953).

18 Hackborn, R., "An automatic network analyzer system", *Microwave J.*, **11**, p. 45 (1968).

19 Shurmer, H. V., "Correction of a Smith-chart display through bilinear transformations", *Electron. Letters*, **5**, p. 209 (1969).

20 Hyltin, T. M., "Microwave integrated electronics for radar and communication systems", *Microwave J.*, **11**, No. 2, p. 51 (1968).

21 Napoli, L., and Hughes, J., "Low-noise integrated X-band receiver", *ibid.*, **11**, No. 7, p. 37 (1968).

22 Arnold, R. D., Bichara, M. R. E., Eberle, J. W., and Reppert, L. M., "Microwave integrated circuit applications to radar systems", *ibid.*, p. 45.

23 Kramer, A., Kelley, D., Solomon, A., Berkovits, G., and Scherer, F., "An integrated X-band c.w. radar front end", *ibid.*, p. 55.

24 Botka, A., Bunker, J., and Gilden, M., "Integrated X-band radar receiver front end", *ibid.*, p. 65.

25 Cornbleet, S., "Microwave solid-state activity in the United Kingdom", *ibid.*, **11**, No. 2, p. 45 (1968).

26 Bowness, C., *Electron Engng.*, p. 2 (Jan., 1968).

27 *Electron. Design* (12 Apr., 1967).

Index

Admittance—
 of biased detector, 16–17
 matrix, 21–2
Ambipolar diffusion coefficient, 138
Applications—
 of integrated circuits, 215–17
 of p-i-n diodes, 143–9
 of transistors, 159
 of varactors, 65–79
Attenuation coefficient, 213–14
Available noise power, 37
Avalanche—
 diode oscillators, 184–202
 frequency, 192
 multiplication, 188

Backward diodes, 123–33
 dynamic range of, 130
Balanced p-i-n diode systems, 146
Band gap, 50
Band structure of GaAs, 167–9
Bandwidth—
 detectors, 14–17
 negative-resistance amplifiers, 73
 p-i-n diode switches, 140, 145–6
Barrier—
 height, 50
 metal/semiconductor junction, 4–8
 p-n junction, 48–56
Beam-lead devices, 208
Bias, effect on capacitance, 58–9
Biased detectors, 13–17
Breakdown voltage, 80, 136
Bridgman method, 178–9
Broad-band detectors, 14
Burnout, 17, 36, 39, 107–8, 128, 132

Capacitance—
 barrier, 28, 64–5
 non-linearity factor of, 46

Capacitance—(*contd.*)
 overhang, 104
 response to applied voltage, 56–8
Cascading of multipliers, 76
Cavity—
 for Read diode, 189
 for tunnel diode, 121–2
Central valley, GaAs, 167–8
Chain multipliers, 76
Characteristic impedance, 206
Circulator, 71, 110, 216
Common-base arrangement, 160–1
Common-emitter arrangement, 160–1
Conductance of barrier, 28, 64–5
Conduction band, 50
Constant-impedance attenuator, 149
Conversion loss, 19, 22–8, 38, 101–2, 104
Coupled-line hybrid circuit, 146
Crystal diode, 1–43
Current gain, 151
Current sensitivity, 9, 128–30
Cut-off frequency—
 transistors, 151
 tunnel diodes, 114
 varactors, 46, 99–100, 104
Cut-off transition time, 62
Czochralski's method, 178–9

D.C. characteristic—
 backward diodes, 124, 127, 129
 crystal diodes, 2
 Gunn diodes, 170
 Schottky-barrier diodes, 95–6
 tunnel diodes, 109–10
Decay time, 62
Density, effective, 113
Depletion capacitance, 48, 52, 59
Depletion layer, 6
 response of boundary, 56–8

219

Index

Depletion width, 52
Detectors—
  general, 8–17, 125–30
  biased, 13–14
Dielectric relaxation time, 56
Diffusion—
  capacitance, 48
  of impurities, 53
  length, 137
  profile, 54
Dimensional requirements of Gunn diodes, 177
Dissipation factor, of strip-line, 210
Domain—
  stable, 169
  travelling, 164–6
Doping/frequency ratio, 172

Effective densities of states, 49
Efficiency—
  frequency multiplier, 76
  Gunn oscillators, 175–7
  Read diodes, 192
Electric field—
  abrupt junction, 52–3
  alloyed-diffused junction, 55–6
  built-in, 81–2
  critical value of, 164, 190
  diffused junction, 54
  Gunn diode, 169–71
  p-n junction, 52–6
  Read diode, 187
Electron potential—
  metal/semiconductor junctions, 5–6
  p-n junctions, 49–56
  transistors, 150
Electron transfer, 164–5
Epitaxy, 44
Epitaxial layers, 3
Equivalent circuit—
  coaxial detector, 16
  illustrating Manley–Rowe equations, 68
  impedance generator, 77
  multiplying chain, 75
  negative-resistance amplifier, 71
  point-contact device, 2
  transistor, 150, 154

Equivalent circuit—(*contd.*)
  tunnel diode, 109–10
  varactor, 44–6

Figure of merit—
  detectors, 10
  mixer material, 26
  transistors, 154
  tunnel diodes, 117
  varactor material, 48
  varactors, 46
Field-effect transistor, 153
Filling fraction, for microstrip, 205–6
Forbidden band, 112
Fourier expansion—
  mixer admittance, 22
  mixer conductance, 28
Frequencies—
  in mixing, 20
  in parametric amplifiers, 67
Frequency—
  conversion, 20–28
  multipliers, 74–7
  of avalanche diodes, 188, 190
  of Gunn diodes, 166

Gain bandwidth product, 74, 117
Gain—
  parametric amplifier, 69–70, 73
  tunnel diode amplifier, 116–7
Gallium-arsenide material preparation, 178–80
Gettering, 212
Gunn diodes—
  dimensional requirements, 177
  dynamic characteristic, 170
  negative mobility, 166, 168
  relaxation effects, 166
Gunn effect devices, 164–83

Harmonic generators, 74
Harmonic multipliers, 47
High-impedance state, of p-i-n diodes, 135–7
Hole-electron pairs, rate of generation, 188
Hybrid integrated circuit, 207–10

220

Index

I.F. impedance, 38
Impedance-transforming networks, 155
Impurity distribution—
　in varactors, 53–6
　in Schottky-barrier diodes, 96–8
Injected holes, 150
Insertion loss, of *p-i-n* diodes, 144–6
Integrated circuits, 203–18
Interdigited structures, 154–5, 159

Junction—
　abrupt, 48–53
　linear-graded, 53–5
　metal/semiconductor, 4–8
　other forms, 55–6
　parameters, of crystal diode detectors, 11
　p-n, 48–56
　tunnelling, 111–14

Kirk effect, 156

Leadless inverted devices (LID), 209
Limited space-charge accumulation (LSA), 167, 171–2
Local tuning, 146
Losses in microstrip, 206, 211
Low-impedance state, of *p-i-n* diodes, 137–40

Magic-T switch, 147
Manley–Rowe equations, 68–9
Manufacture—
　avalanche diodes, 198–200
　backward diodes, 132–3
　crystal diodes, 34–5
　Gunn diodes, 178–80
　Schottky barrier diodes, 105–8
　transistors, 158–9
　tunnel diodes, 119–20
　varactors, 79–82
Mass, effective, 50, 113
Maxwell–Boltzman law, 7
Measurements—
　backward diodes, 133
　crystal diodes, 35–42
　integrated circuits, 212–15

Measurements—(*contd.*)
　overall noise factor, 40
　Schottky-barrier diodes, 108
　transistors, 161–2
　tunnel diodes, 122
　varactors, 82–90
Microelectronics, 204
Microstrip, 204–10
Mixer—
　crystal diode, 17–34
　Schottky-barrier diode, 101–4
　backward diode, 130–2
Modulators, 143–7
Molecular Electronics for Radar Applications (MERA), 215–6
Monolithic circuits, 210–12
Mount—
　crystal detector, 16
　Gunn diode, 181
　p-i-n diode, 146
　tunnel diode, 120–2
　varactor, 80

Negative Q-factor of Read diode oscillators, 189
Negative resistance amplifier, 70–4
Negative resistance, of tunnel diodes, 114–16
Noise—
　bandwidth, 10
　equivalent resistance, 10
　factor, overall, 17–19, 36–7, 39, 40, 130–2
　low-frequency (flicker), 30, 40–2, 107–8, 131–2
　microwave, 32
　ratio, 19, 28–33, 38
　shot, 29
　temperature, effective, 70
　thermal, or Johnson, 29
　tube, 37
　under local-oscillator excitation, 32–3
　voltage, 10
n-p-i-p oscillator (*see* Read diode)

Parametric amplifiers—
　degenerate, 65–7

Index

Parametric amplifiers—(*contd.*)
  general, 47, 65–74
  negative-resistance, 70–4
  non-degenerate, 67–8
Permittivity, effective relative, 206
Phase shifters, 143, 148
Phased-array radars, 215–6
Photo-etching, 209
p-i-n—
  diodes, 134–49
  oscillators, 194–6
Planar techniques for transistors, 153
Poisson's law, 51, 54
Power capability—
  Gunn diodes, 176
  p-i-n diode oscillators, 196
  p-i-n diode switches, 142
  p-n diode oscillators, 193–4
  Read diodes, 185, 189–90
  transistors, 159
  varactor multipliers, 76
Power limiter—
  p-i-n, 143–4
  varactor, 78–9
Pump frequency, 67
"Punch-through" varactors, 63–4

Q-factor—
  chart for evaluating, 90
  p-i-n diode, 135, 140
  Read diode, 189, 191
  transmission line resonator, 207
  varactors, 47, 64–5, 85–90

Radars, phased array, 215–16
Read diode, 184, 186–92
Receivers—
  general, 1
  superheterodyne, 17–19
  three stage, 19
  video, 8–11
Recombination of charge carriers, 137
Resistance, spreading, 2

Scattering mechanisms, 168–9
Scattering parameters, 161–2
Sandwich structure, 180

Satellite valley, 167–8
Schottky-barrier diodes, 93–108
Self-resonant frequency, 114
Sensitivity—
  limiting, 11
  tangential, 11, 35–6, 100–1, 104, 126
Shape ratio, microstrip, 205
Shockley diffusion current, 138
Signal/noise ratio, 10–11
Skin effect, in *p-i-n* diodes, 139
"Snap-off" diodes, 47, 59–63
Space charge, 6, 51, 172
Spreading resistance, 2
Square wave modulator, 143–7
State of the art—
  avalanche diodes, 186
  Gunn diodes, 177
"Step-recovery" diodes, 47, 59–63
"Stick" diodes, 146
Stoichiometric junction, 58
Storage capacitance, 58–9
Storage phase, 60–1
Stored charge in *p-i-n* diodes, 141–2
Stray capacitance, 82, 86
Strip-line laminates, 207, 209
Strip transmission lines, 203–7
Superheterodyne system, 1
Surface leakage, 136
Surface-oriented devices, 216
Surface states, 8
Switches, 143, 147–8
Switching—
  by varactors, 77–8
  ratio, 140
  speed, 141–2

Thermionic work function, 4–5
Transistors—
  bipolar, 53–7
  field-effect, 153
  interdigited, 159
  maximum frequency of oscillation, 155
  mesa, 152
  metal-oxide-semiconductor, 153
  micro-alloy, 152
  surface barrier, 152
  unipolar, 153

Index

Transition phase, 61
Transmission-line analogues, 160
Transmission measurements, 83
Trapped plasma resonance mode (TRAPATT), 185
Travelling domains, 164–6, 169
Triplate, 204
T–R switches, 216
Tuning—
    of avalanche diodes, 185
    of Gunn diodes, 167, 173–4
    of transistors, 161
    varactors, 78
Tunnel diodes—
    amplifiers, 116–18
    oscillators, 18–19
    characteristic equation, 115
    general, 109–23
    stability of, 115–16
Tunnelling current, 113
Turn-off time, 142

Unwanted modes, 121

Vacuum deposition, 209
Valence band, 50
Varactor—
    diodes, 44–92
    limiter, 78–9
    "punch through", 63–4
    switching, 77–8
Velocity of charge carriers in an electric field—
    for Gunn diodes, 181
    near a p-n junction, 56–8
Video receiving system, 1
Video resistance, 35
Voltage, breakdown, 80, 136
Voltage controlled attenuators, 143, 148–9
Voltage standing wave ratio measurements, 36–9
Voltage stress at microwave frequencies 137
Voltage-tunable Gunn oscillator, 181

Wavelength in microstrip, 206, 213
Whiskers—
    for backward diodes, 125, 132–3
    for crystal diodes, 3, 34

Y-factor, 19

Zener effect, 190